STREAMLINING SPACE LAUNCH RANGE SAFETY

Committee on Space Launch Range Safety

Aeronautics and Space Engineering Board

Commission on Engineering and Technical Systems

National Research Council

NATIONAL ACADEMY PRESS
Washington, D.C.

NOTICE: The project that is the subject of this report was approved by the Governing Board of the National Research Council, whose members are drawn from the councils of the National Academy of Sciences, the National Academy of Engineering, and the Institute of Medicine. The members of the committee responsible for the report were chosen for their special competencies and with regard for appropriate balance.

This study was supported by the U.S. Air Force Space Command under contract No. FO5604-99-C-9004. Any opinions, findings, conclusions, or recommendations expressed in this publication are those of the authors and do not necessarily reflect the view of the Air Force.

International Standard Book Number: 0-309-06931-9

Available in limited supply from: Aeronautics and Space Engineering Board, HA 292, 2101 Constitution Avenue, N.W., Washington, DC 20418. (202) 334-2855 *www4.national-academies.org/cets/asebhome.nsf*

Additional copies available for sale from: National Academy Press, 2101 Constitution Avenue, N.W. Box 285, Washington, DC 20055. 1-800-624-6242 or (202) 334-3313 (in the Washington Metropolitan area). *www.nap.edu*

Copyright 2000 by the National Academy of Sciences. All rights reserved.

Printed in the United States of America

THE NATIONAL ACADEMIES

National Academy of Sciences
National Academy of Engineering
Institute of Medicine
National Research Council

The **National Academy of Sciences** is a private, nonprofit, self-perpetuating society of distinguished scholars engaged in scientific and engineering research, dedicated to the furtherance of science and technology and to their use for the general welfare. Upon the authority of the charter granted to it by the Congress in 1863, the Academy has a mandate that requires it to advise the federal government on scientific and technical matters. Dr. Bruce M. Alberts is president of the National Academy of Sciences.

The **National Academy of Engineering** was established in 1964, under the charter of the National Academy of Sciences, as a parallel organization of outstanding engineers. It is autonomous in its administration and in the selection of its members, sharing with the National Academy of Sciences the responsibility for advising the federal government. The National Academy of Engineering also sponsors engineering programs aimed at meeting national needs, encourages education and research, and recognizes the superior achievements of engineers. Dr. William A. Wulf is president of the National Academy of Engineering.

The **Institute of Medicine** was established in 1970 by the National Academy of Sciences to secure the services of eminent members of appropriate professions in the examination of policy matters pertaining to the health of the public. The Institute acts under the responsibility given to the National Academy of Sciences by its congressional charter to be an adviser to the federal government and, upon its own initiative, to identify issues of medical care, research, and education. Dr. Kenneth I. Shine is president of the Institute of Medicine.

The **National Research Council** was organized by the National Academy of Sciences in 1916 to associate the broad community of science and technology with the Academy's purposes of furthering knowledge and advising the federal government. Functioning in accordance with general policies determined by the Academy, the Council has become the principal operating agency of both the National Academy of Sciences and the National Academy of Engineering in providing services to the government, the public, and the scientific and engineering communities. The Council is administered jointly by both Academies and the Institute of Medicine. Dr. Bruce M. Alberts and Dr. William A. Wulf are chairman and vice chairman, respectively, of the National Research Council.

COMMITTEE ON SPACE LAUNCH RANGE SAFETY

ROBERT E. WHITEHEAD, *chair*, National Aeronautics and Space Administration (retired), Henrico, North Carolina
W. GAINEY BEST II, Lockheed Martin Astronautics, Denver, Colorado
JOHN L. BYRON, Johnson Controls, Inc., Cocoa Beach, Florida
BENJAMIN A. COSGROVE, Boeing Commercial Airplane Group (retired), Seattle, Washington
JAMES W. DANAHER, National Transportation Safety Board (retired), Alexandria, Virginia
KINGSTON A. GEORGE, aerospace consultant, Santa Maria, California
BILL HAWLEY, Hughes Space and Communications, Los Angeles, California
JAMES K. KUCHAR, Massachusetts Institute of Technology, Cambridge
JOYCE A. McDEVITT, Futron Corporation, Washington, D.C.
JOSEPH MELTZER, Aerospace Corporation (retired), Redondo Beach, California
JIMMEY MORRELL, U.S. Air Force (retired), Melbourne, Florida
NORMAN H. SCHUTZBERGER, TRW Components International, Torrance, California

Liaison from the Aeronautics and Space Engineering Board

FREDERICK HAUCK, AXA Space, Bethesda, Maryland

Staff

ALAN ANGLEMAN, Study Director
DOUGLAS BENNETT, Research Associate
CHRIS JONES, Senior Administrative Assistant
GEORGE LEVIN, Director, Aeronautics and Space Engineering Board
JENNIFER PINKERMAN, Research Associate
LINDA VOSS, Technical Writer
MARVIN WEEKS, Senior Administrative Assistant

AERONAUTICS AND SPACE ENGINEERING BOARD

WILLIAM W. HOOVER, *chair*, U.S. Air Force (retired), Williamsburg, Virginia
A. DWIGHT ABBOTT, Aerospace Corporation, Los Angeles, California
WILLIAM F. BALLHAUS, JR., Lockheed Martin Corporation, Bethesda, Maryland
RUZENA BAJSCY, NAE, IOM, University of Pennsylvania, Philadelphia
ANTHNY J. BRODERICK, Aviation Safety Consultant, Catlett, Virginia
AARON COHEN, NAE, Texas A&M University, College Station
DONALD L. CROMER, U.S. Air Force (retired), Lompoc, California
HOYT DAVIDSON, Donaldson, Lufkin, and Jenrette, New York, New York
ROBERT A. DAVIS, The Boeing Company (retired), Seattle, Washington
DONALD C. FRASER, NAE, Boston University, Boston, Massachusetts
JOSEPH FULLER, JR., Futron Corporation, Bethesda, Maryland
ROBERT C. GOETZ, Lockheed Martin Skunk Works, Palmdale, California
RICHARD GOLASZEWSKI, GRA Inc., Jenkintown, Pennsylvania
JAMES M. GUYETTE, Rolls-Royce North America, Reston, Virginia
FREDERICK HAUCK, AXA Space, Bethesda, Maryland
JOHN K. LAUBER, Airbus Industrie of North America, Washington, D.C.
GEORGE MUELLNER, The Boeing Company, Seal Beach, California
DAVA J. NEWMAN, Massachusetts Institute of Technology, Cambridge
JAMES G. O'CONNOR, NAE, Pratt & Whitney (retired), Coventry, Connecticut
WINSTON E. SCOTT, Florida State University, Tallahassee
KATHRYN C. THORNTON, University of Virginia, Charlottesville
DIANNE S. WILEY, Northrop Grumman, Pico Rivera, California
RAY A. WILLIAMSON, George Washington University, Washington, D.C.

Staff

GEORGE LEVIN, Director

Preface

Space launch is inherently risky, and accidents are not uncommon. However, the U.S. national ranges have an outstanding safety record. Never has a member of the public or the launch site workforce been killed as a result of a launch from the national ranges. The aging systems that have achieved this record are being modernized to improve performance and reduce costs. As part of this effort, the National Research Council (NRC) was asked to determine if alternate approaches to public safety might be more efficient and less expensive than current methods.

Because space launch activities and associated safety practices are highly technical, this study examined the technologies associated with launch range safety and assessed the ability of advanced technologies to improve efficiency and reduce costs. However, because safety also depends on other factors, the study was not a purely technological assessment. The statement of task called for a comprehensive review that included range safety guidelines and procedures. In addition, during the course of the study the committee concluded that a complete response to the statement of task would require that top-level organizational issues related to the efficiency and cost of range safety also be examined. The Air Force Space Command, which sponsored this study, concurred, and the committee's findings and recommendations are framed accordingly.

The NRC appointed 10 individuals to the Committee on Space Launch Range Safety, which conducted this study. Like all NRC study committees, the membership was announced and comments from the general public were solicited regarding the committee's composition and balance. A number of comments urged that the committee be expanded to include individuals who had worked for range safety organizations. In response, the NRC decided to add two members: a former commander of the 45th Space Wing, which operates the Eastern Range, and a former chief engineer and deputy director for safety at the 30th Space Wing, which operates the Western Range. Some committee members had experience with major launch vehicle manufacturers, satellite manufacturers, and other users, as well as technical expertise in risk analysis, global positioning system (GPS) technology, and public safety. Also, to provide an impartial, outside perspective, several committee members had little or no launch industry experience. Thus, the committee was well qualified to conduct both the technical and nontechnical aspects of the statement of task.

This study benefited from an extraordinary level of public interest. More than 100 individuals from interested organizations and members of the general public attended the committee's information-gathering meetings, which included opportunities for public input. This broad participation greatly contributed to the committee's deliberations, and the committee is indebted to everyone who gave of their time and talent during the meetings.

This report is being issued in parallel with a number of other reports concerned with launch range safety, infrastructure, operations, and organization. The findings and recommendations herein endorse some of the actions currently under way, recommend the acceleration and extension of others, and suggest some new initiatives. In particular, the committee recommends that the Air Force retain its key safety standards and make greater use of those standards for managing risk. By moving away from costly risk avoidance practices, the Air Force would conform range safety procedures to accepted risk standards and reduce costs for both the Air Force and the user community without compromising public safety.

This report has been reviewed by individuals chosen for their diverse perspectives and technical expertise, in accordance with procedures approved by the NRC's Report Review Committee. The purpose of this independent review is to provide candid and critical comments that will assist the authors and the NRC in making the published report as sound as possible and to ensure that the report meets institutional standards for objectivity, evidence, and responsiveness to the study charge. The content of the review comments and draft manuscript remain confidential to protect the integrity of the deliberative process. We wish to thank the following individuals for their participation in the review of this report:

Silas Baker, Jr., Lockheed Martin (retired)
Robert Crippen, Thiokol Propulsion
Donald Cromer, U.S. Air Force (retired)
Robert Frosch, Harvard University
Daniel Hastings, Massachusetts Institute of Technology
Donald Henderson, U.S. Air Force (retired)
Hal Lewis, University of California, Santa Barbara
James Means, SRI International
Sheila Widnall, Massachusetts Institute of Technology

While the individuals listed above have provided many constructive comments and suggestions, responsibility for the final content of this report rests solely with the authoring committee and the NRC.

Robert E. Whitehead, *Chair*
Committee on Space Launch Range Safety

Contents

EXECUTIVE SUMMARY 1

1 INTRODUCTION 7
Objectives, 7
Study Processes and Approach, 7
Organization of This Report, 8
References, 8

2 BACKGROUND 9
National Space Launch Policy, 9
Roles and Responsibilities of the Ranges and Users, 10
Safety Standards, 12
Commercial Cost Drivers, 12
References, 13

3 RISK MANAGEMENT APPROACHES TO SAFETY 14
Philosophy of EWR 127-1, 14
Roles and Responsibilities of the Air Force Space Command and Air Force Materiel
  Command, 15
Risk Criteria, Risk Management, and Analysis Methods, 18
References, 25

4 FLIGHT SAFETY REQUIREMENTS 26
Tracking, 26
Telemetry, 27
GPS Metric Tracking, 29
Range Modernization, 31
Autonomous Flight Termination Systems with GPS, 33
Reusable Launch Vehicles, 33
References, 35

5 INCURSIONS 36
Current Guidelines and Procedures, 36
Planned Improvements and Additional Recommendations, 40
References, 41

APPENDIXES

A FINDINGS AND RECOMMENDATIONS   43
B BIOGRAPHIES OF COMMITTEE MEMBERS   46
C PARTICIPANTS IN COMMITTEE MEETINGS   49
D STUDIES RELATED TO SPACE LAUNCH RANGE SAFETY   51
E SAFETY MODELING AND ANALYSIS   53

ACRONYMS   57

Tables and Figures

TABLES

3-1 Comparison of Maximum Acceptable Collective Risks, 19
3-2 Probability of Failure vs. Phase for the Atlas IIAS, 23

FIGURES

ES-1 Comparison of life-cycle costs for radar and GPS-based range tracking systems, 4

3-1 Air Force roles and responsibilities for space launch, 16
3-2 Instantaneous impact point trace and Africa gate location for Titan IV-B25, 21
3-3 Ground track and elevation angle for an Atlas IIA launched from Pad 36A at the Eastern Range on an initial flight azimuth of 104 degrees, 22
3-4 Probability of failure vs. phase for the Atlas IIAS, 23

4-1 Changes in range tracking support under the RSA range modernization program, 28
4-2 Comparison of life-cycle costs for radar and GPS-based range tracking systems, 32
4-3 Flight safety angle limits, 34

5-1 Samples of multiple boat-hit contours, 39

Executive Summary

INTRODUCTION

The U.S. space program is rapidly changing from an activity driven by federal government launches to one driven by commercial launches. In 1997, for the first time commercial launches outnumbered government launches at the Eastern Range (ER), located at Cape Canaveral Air Station, Florida. Commercial activity is also increasing at the Western Range (WR), located at Vandenberg Air Force Base, California. The government itself is emulating commercial customers, shifting from direct management of launch programs to the purchase of space launch services from U.S. commercial launch companies in an open, competitive market.

The fundamental goal of the U.S. space program is to ensure safe, reliable, and affordable access to space. Despite the inherent danger of space launches, the U.S. space program has demonstrated its ability to protect the public. No launch site worker or member of the general public has been killed or seriously injured in any of the 4,600 launches conducted at the ER and WR during the entire 50-year history of the space age.

Reliability and affordability have been more difficult to achieve. As the federal government relies more on the commercial space sector to launch government payloads, the vitality, viability, and global competitiveness of the U.S. commercial launch industry are becoming increasingly important. Because range safety costs are an important element of total launch costs, it would be beneficial to streamline safety processes without lowering current safety standards. This study responds to a request from the Air Force Space Command (AFSPC), which operates the ER and WR, to determine if range safety processes can be made more efficient and less costly without compromising public safety. This summary presents six primary recommendations, which address risk management, Africa gates, roles and responsibilities, range safety documentation (i.e., *Eastern and Western Range Safety Requirements* [EWR 127-1]),[1] global positioning system (GPS) receiver tracking systems, and risk standards for aircraft and ships. The main body of the report contains eight other recommendations that would make smaller contributions to achieving the study goals. The report also contains 14 findings that support the recommendations and state the committee's conclusions in areas where the committee decided recommendations were not warranted.

COMMITTEE TASK

The statement of task for this study specified three areas of interest:

- a top-level, independent review of the Air Force's safety guidelines and procedures for government and commercial space launches as published in EWR 127-1 to determine if there are alternative approaches to the protection of the general public that are both more efficient and less expensive
- an independent assessment of the current and planned range safety and flight termination systems and procedures for government and commercial space launches to estimate the technical feasibility as well as the cost effectiveness of an autonomous GPS flight termination system
- an independent examination of the Air Force's safety guidelines and procedures associated with incursions of aircraft and ships into restricted air space and waters to determine if holds and delays of government and commercial space launches can be reduced while still maintaining an acceptable level of safety

---

[1] EWR 127-1 is the primary range safety requirements document for both the ER and WR.

RISK CRITERIA AND RISK MANAGEMENT

With any endeavor, it is generally desirable to increase return or value and reduce risk. This often involves defining an acceptable risk level as a standard to which risk then can be managed. Once this standard has been met, the venture may be considered safe. The fundamental analytical risk standard used by the WR and ER for collective risk to the general public is expressed in terms of *casualty expectation* (E_c). For each launch, E_c must be less than 30×10^{-6}; at a rate of 33 launches per year, this is equivalent to one serious injury or fatality every 1,000 years. The committee considered recommending different risk standards for collective risk and individual risk, P_c (discussed below). The current standards, however, are in line with the level of risk characteristic of many other fields, domestically and internationally, in which the public is involuntarily exposed to risk. Also, the committee determined that the efficiency of range operations could be significantly improved without lowering safety standards and that higher standards are not needed to protect the public. Therefore, the committee supports the continued use of 30×10^{-6} as the collective risk standard for space launches at the ER and WR.

A recurring theme in the findings and recommendations of this report is the importance of managing risk to the accepted standards. Risk management ensures that launch vehicles are manufactured and launch operations are conducted to achieve established safety standards. Risk management also allows weighing the costs and benefits of alternative approaches for meeting risk standards. Currently, commercial operators must comply with federal range safety requirements that are implemented in a way that leads to risk avoidance instead of risk management. The correct goal, however, would be to meet established safety standards in a cost-effective manner that facilitates planned operations, rather than reducing risk to the lowest possible level regardless of the costs or requiring the most conservative application of risk standards throughout the range safety process.

Primary Recommendation on Risk Management. AFSPC should define objective, consistent risk standards (e.g., casualty expectation, E_c, of 30×10^{-6}, and individual risk, P_c, of 1×10^{-6}) and use them as the basis for range safety decisions. Safety procedures based on risk avoidance should be replaced with procedures consistent with the risk management philosophy specified by EWR 127-1. Destruct lines and flight termination system requirements should be defined and implemented in a way that is directly traceable to accepted risk standards.

AFRICA GATES

Because a launch vehicle may pass over populated landmasses before orbital insertion, strict limits are often provided in the form of "gates" in the impact limit lines (ILLs) and destruct lines that define the range of allowable flight paths. If the vehicle does not pass through the gate, the flight is terminated. The ILLs, destruct lines, and gates are designed to ensure that debris returns to Earth more than 50 miles from the coasts of major populated landmasses. At the ER, the downrange location of gates and destruct lines as well as requirements for downrange coverage by flight termination, telemetry, and tracking systems, are not directly related to accepted risk standards (e.g., E_c of 30×10^{-6} or P_c of 1×10^{-6}) but to a risk-avoidance policy that discourages the overflight of inhabited landmasses "whenever possible" (EWR 127-1, paragraph 2.3.6). The committee recognizes that avoiding inhabited landmasses is often the best approach for meeting risk standards. However, using risk standards to evaluate alternate approaches is more rigorous than relying on subjective criteria, such as "whenever possible."

The positioning of gates is a function of the launch vehicle, flight azimuth, and location of inhabited landmasses. The Africa gates are typically beyond the range of uprange radar tracking facilities and require the use of downrange facilities on the islands of Antigua and Ascension. Moving the Africa gates uprange could reduce the cost of safety-related assets, decrease the complexity of range safety operations, and reduce holds and delays. Based on historical failure data and reliability requirements, moving the Africa gates to within the reach of uprange flight termination systems (FTSs) and tracking systems is unlikely to increase E_c significantly or violate established limits. In addition, the committee knows of no international agreements that would preclude moving the gates uprange. Thus, in terms of range safety there is no clear justification for retaining downrange assets at Antigua and Ascension. It may also be feasible to move other gates uprange and further reduce the need for downrange facilities at the ER. The WR already avoids the use of downrange flight termination, telemetry, and tracking systems by constraining allowable azimuths of orbital launches during the uprange portion of flights to avoid flying over populated areas.

Primary Recommendation on Africa Gates. While other requirements may exist, from the perspective of launch range safety the Air Force should move the Africa gates to within the limits of uprange flight termination and tracking systems; eliminate the use of assets in Antigua and Ascension for range safety support; and conduct a detailed technical assessment to validate the feasibility of moving other gates uprange. If other requirements for downrange tracking exist, AFSPC should validate those requirements and reexamine this recommendation in light of the additional requirements.

ROLES AND RESPONSIBILITIES

AFSPC has transferred to the Air Force Materiel Command (AFMC) responsibility for development, developmental testing and evaluation, and sustaining engineering of range safety ground systems. Organizational responsibilities for many other range safety processes and procedures, however, are still inconsistent with the current memorandum of agreement between AFSPC and AFMC on spacelift roles and responsibilities. In addition to the operational workforce, each AFSPC range safety office also has an engineering workforce that establishes flight safety system design and testing requirements and certifies that flight safety systems meet safety requirements at the component, subsystem, and system levels. These acquisition-like functions overlap the responsibilities of AFMC.

If properly executed, the complete transfer of range safety development, developmental testing and evaluation, and sustaining engineering to AFMC would increase efficiency and reduce costs without compromising safety by eliminating overlapping responsibilities between the ranges and AFMC, by minimizing differences in range safety policies and procedures applicable to the WR and ER, and by allowing users to deal with a single office when seeking approval to use new or modified systems on both ranges. This transfer could be facilitated by issuing an Air Force Instruction describing the certification of flight safety systems for commercial, civil, and military launches at the ER or WR. The instruction should also describe interfaces among responsible organizations, such as AFSPC, AFMC, the Federal Aviation Administration (FAA), the National Aeronautics and Space Administration, and commercial contractors.

Primary Recommendation on Roles and Responsibilities. The Air Force should fully implement the memorandum of agreement between AFSPC and AFMC on spacelift roles and responsibilities. This would consolidate within AFMC the acquisition-like functions related to safety that are now performed by AFSPC organizations at the Eastern and Western Ranges. These functions include developmental testing and evaluation, sustaining engineering, and certifying that system designs meet safety requirements. To manage the safety aspects of the acquisition-like functions specified in the memorandum of agreement, AFMC should establish an independent safety office. Operational responsibilities, such as generating safety requirements, operational testing and evaluation, and all prelaunch and launch safety operational functions, would be retained by AFSPC.

EWR 127-1

EWR 127-1 is issued jointly the by 30th Space Wing, which operates the WR, and the 45th Space Wing, which operates the ER. EWR 127-1 specifies in detail how to comply with established risk standards rather than expecting users to develop their own methods of compliance. These detailed requirements create the need for extensive "tailoring" of EWR 127-1 for each new launch vehicle to allow the use of alternate solutions that are more practical than the specified methods of compliance. The committee believes that a more effective approach would be to streamline EWR 127-1 to focus on baseline performance-based requirements and move detailed solutions and lessons learned to a range user's handbook. This would reduce or eliminate the need for tailoring and draw a clear distinction between non-negotiable performance-based requirements and recommended methods of compliance that can be waived if an equally effective alternative is available and the user accepts the burden of demonstrating its effectiveness.

Primary Recommendation on EWR 127-1. AFSPC should simplify EWR 127-1 so that all requirements are performance based and consistent with both established risk standards for space launch (e.g., E_c of 30×10^{-6}) and objective industry standards. The process of revising EWR 127-1 should include the following steps:

- Eliminate requirements that cannot be validated.
- Remove all design solutions from EWR 127-1.
- Establish a range user's handbook or other controlled document to capture lessons learned and design solutions recognized by the ranges as acceptable means of compliance. (Requirements should be retained in EWR 127-1.)
- Form a joint government/industry team to establish procedures for periodically updating EWR 127-1 and ensuring that future requirements are performance based.
- Converge the modeling and analysis approaches, tools, assumptions, and operational procedures used at the Western and Eastern Ranges.

GPS FLIGHT ARCHITECTURE

AFSPC plans to implement a GPS-based flight architecture at the ER and WR, which will reduce the cost of upgrading, maintaining, and operating the radar system (see Figure ES-1). A GPS-based tracking system will permit shutting down 11 of the 20 tracking radars currently used to support launch operations at the ER and WR. Three of the remaining radars will be needed only to support launches of the space shuttle.

There are two approaches for implementing a GPS-based tracking system. A GPS translator system would retransmit GPS signals received by a launch vehicle to the ground, where vehicle position and velocity would be computed. This approach would have high bandwidth requirements for communications signals sent from the launch vehicle to ground stations and from ground stations to GPS processor sites.

The alternative would be to use GPS receivers on each vehicle to calculate vehicle position and velocity data, which would then be transmitted to the ground. A GPS receiver system would have low bandwidth requirements and enable an open system architecture compatible with future concepts, such as space-based ranges and autonomous or semiautonomous

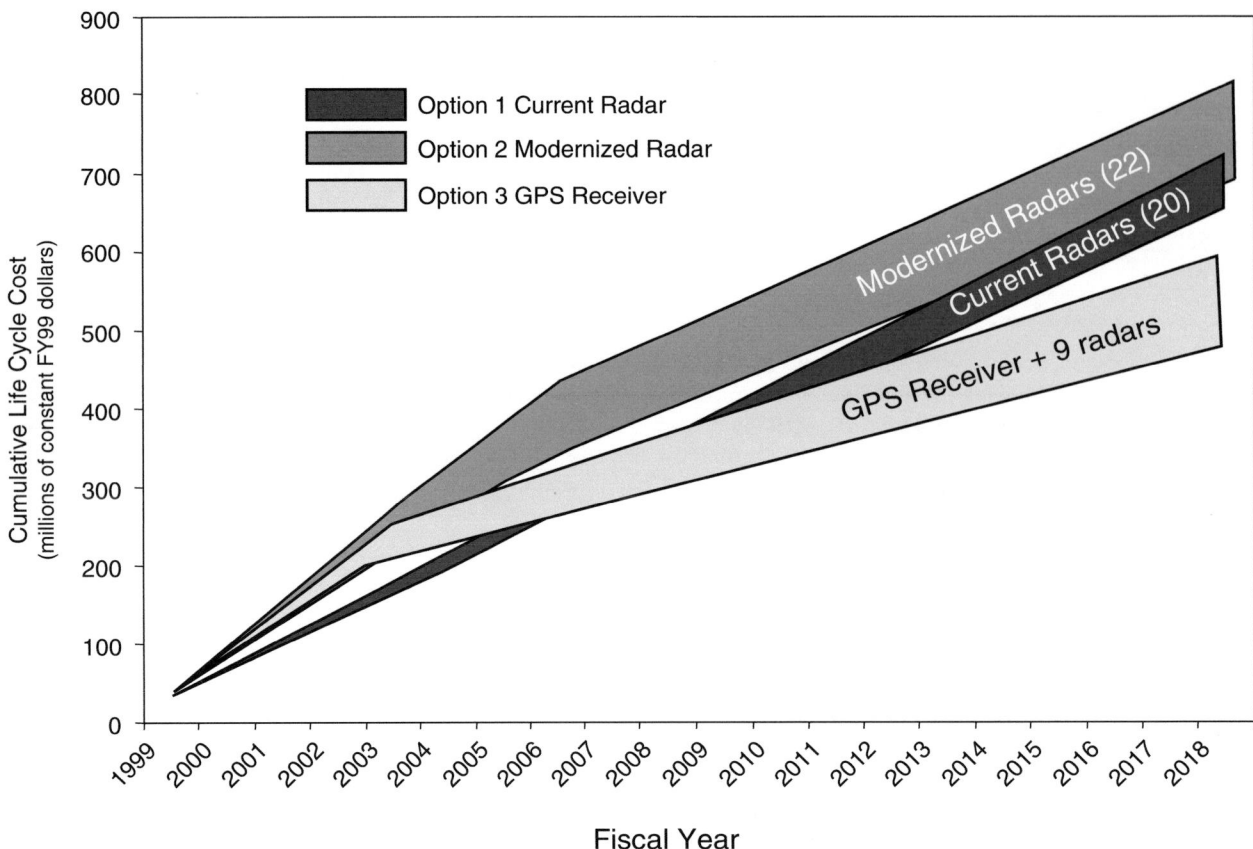

FIGURE ES-1 Comparison of life-cycle costs for radar and GPS-based range tracking systems. Source: Finn and Woods, 1999.

FTSs. Autonomous features have already been implemented in current FTSs in the form of inadvertent separation destruct systems, which sense the onset of unplanned vehicle breakups and, in many cases, automatically initiate flight termination. A semiautonomous system could be developed in which the uprange portion of flight is monitored using traditional human-in-the-loop FTS procedures. Then, as the vehicle travels downrange and the risk profile decreases, the FTS could be shifted to a fully autonomous mode. This has the potential to reduce costs, improve responsiveness to unplanned events, and enable ranges to more easily support a broad complement of launch vehicles and mission profiles.

With the incorporation of onboard GPS receivers, fully autonomous FTSs would become technically feasible, but additional research and testing is needed to resolve outstanding issues related to system performance requirements, development and validation costs, and public acceptability. The successful deployment of semiautonomous systems, which would provide operational benefits even if a fully autonomous system is never developed, would help resolve these issues.

Finding. For space launches, an onboard GPS receiver tracking system would be more versatile and have lower total life-cycle costs than GPS translator or radar tracking systems.

Primary Recommendation on GPS Receivers. AFSPC should deploy a GPS receiver tracking system as the baseline range tracking system for space launch vehicles. The transition to GPS-based tracking should be completed as rapidly as feasible.

MARINE AND AIRCRAFT INCURSIONS

Aircraft and marine incursions into restricted airspace and waters have contributed to only a small percentage of launch holds and scrubs at either the ER or WR. However, when they do occur, these delays can be highly disruptive and costly, for both the range and the user. Also, increases in marine and air traffic near the launch area and more frequent space launches are expected to increase the number of boat and aircraft intruders, especially at the ER. An improved launch communications and notification process would benefit the general public, the Air Force, and range users. Options include making greater use of public media, such as newspapers, radio and television broadcasts, the Internet, notices at public marinas and general aviation airports, and aviation and marine weather broadcasts; reviewing the adequacy of current signs, lights, and other warning devices at marinas and along the coast; and modifying warning devices to increase their effectiveness in deterring marine incursions.

Improving the notification process alone, however, will not completely solve the intruder problem. The committee recommends immediate improvements so that surface and aircraft intruders can be detected earlier and cleared from the launch area more quickly. These improvements should include the use of commercial aircraft equipped with suitable surveillance, navigation, communications, and image recording systems for marine intruders and surveillance systems for aircraft intruders.

AFSPC should also aggressively enforce restrictions against intruders at both ranges to encourage compliance with launch notifications. In cooperation with the U.S. Coast Guard, the FAA, the U.S. Attorney's Office, and other regulatory and law enforcement agencies, AFSPC should initiate administrative and regulatory changes to facilitate enforcement action against intruders who were afforded ample, timely launch notifications.

All of the actions described above are based on the establishment of hazardous launch areas (e.g., flight hazard and flight caution areas), which extend downrange from the launch site along the intended flight azimuth. The size and shape of these areas are based on calculations of the probability, P_i, of hitting an individual ship or aircraft. The calculations take into account the characteristics of specific launch vehicles and payloads, failure modes and effects (including toxic hazards), and weather considerations.

The individual risk standard for members of the general public is 1×10^{-6}. This means that the probability, P_c, that a member of the public at any particular place will be killed or seriously injured shall not exceed 1×10^{-6} for any launch. A different risk standard is appropriate for individual ship-hit probability because hitting a ship with a piece of debris will not necessarily result in casualties.

EWR 127-1 does not specify a risk standard for aircraft incursions. Risk standards are used to manage risk to mission-essential aircraft, but the standards are applied differently at the ER and WR and are not supported by analyses showing that the standards are consistent with other safety criteria used by the ranges. For example, P_i should be calculated differently for aircraft than for ships because even small pieces of debris can endanger aircraft.

The ER and WR use predefined restricted areas to protect public aircraft from launch hazards. These areas are sized to keep aircraft totally away from hazardous operations and are plotted on standard aeronautical charts. Prior to each launch the flying public is warned to remain clear of restricted areas involved in that launch.

If intruder aircraft are in restricted areas prior to launch, the launch could safely proceed if the aircraft will remain clear of the regions of actual hazard. This could be accomplished through the use of buffer zones around each hazard area. The buffer zones should be large enough so that, even if an aircraft outside the buffer zone turns toward the hazard area at the beginning of the launch commit cycle, the aircraft could not reach the hazard area until after the launch vehicle has cleared the area. These buffer zones are not needed for aircraft flying under the direction of air traffic controllers in airways outside the hazard areas.

Primary Recommendation on Risk Standards for Aircraft and Ships. AFSPC should apply the individual ship-hit criterion, P_i, of 1×10^{-5} to the ship exclusion process at the Eastern Range in the same way it is used at the Western Range. EWR 127-1 should be modified to specify an aircraft-hit P_i limit of 1×10^{-6} (properly calculated to include the probability of impact for very small pieces of debris). Prior to each launch, the range should establish aircraft hazard areas (based on the aircraft P_i) and buffer zones (for uncontrolled aircraft in the vicinity of the hazard area). Launches should be allowed to proceed as long as no intruder aircraft are in the hazard area or buffer zone.

SUMMARY REMARKS

Implementation of the committee's recommendations would streamline range safety processes, resulting in substantially lower costs and higher efficiency without compromising safety. The recommendations and associated findings are grounded on the universal application of the Air Force's long-established risk management approach to space launch range safety. Implementation of the recommendations would eliminate the overly cautious risk-avoidance practices that have crept into established range safety practices, reform EWR 127-1 to focus on performance-based requirements based on objective risk standards, create a single range safety office under AFMC to consolidate nonoperational range safety activities, greatly reduce the need for downrange safety facilities, reduce launch holds and scrubs caused by aircraft and ship incursions, and upgrade the ranges with GPS receiver tracking systems to reduce costs and pave the way for long-term improvements, such as semiautonomous FTSs and space-based ranges.

The recommendations in this report are consistent with and complementary to many ongoing efforts to modernize space launch infrastructure and procedures. Some of the recommendations can be implemented immediately, while others must be part of longer term upgrades to the infrastructure. All of them will require cooperation among the ranges, other elements of the Air Force, other government agencies involved in space launches, and range users. If the recommendations are carefully implemented, everyone involved would benefit from a safe, more economical, and more competitive U.S. space launch capability. Together with the results of related studies, the Air Force now has enough information to create timetables, establish priorities, assign responsibilities, and take action to improve U.S. space launch capabilities.

REFERENCES

EWR 127-1 (Eastern and Western Range Safety Requirements). 1997. Available on line at: *http://www.pafb.af.mil/45sw/rangesafety/ewr97.htm* January 20, 2000.

Finn, G., and T. Woods, 1999. Spacelift Range Metric Tracking LCC (Life Cycle Cost) Review: Response to NRC Request. August 10, 1999. Briefing materials prepared by G. Finn, Aerospace Corporation, and T. Woods, Air Force Space and Missile Systems Center, for the Space Launch Range Safety Committee, August 13, 1999.

1

Introduction

As part of its efforts to modernize and streamline space launch operations, the Air Force chartered the National Research Council (NRC) to review safety guidelines and procedures for government and commercial space launches at the national ranges. The need for an independent assessment by the NRC was identified in the *Range Integrated Product Team (IPT) Report* (USAF, 1998), which was prompted by the Commercial Space Industry Leaders' Conference that took place in December 1997. At that conference, the following opportunities for improving U.S. space launch capabilities were defined:

- streamlining processes to reduce operational costs and personnel requirements while preserving public safety
- reexamining policies for customer use of limited launch range resources
- improving support for commercial users
- learning from the experience of foreign launch range operations

The *Range IPT Report* was produced by a task force of representatives from the Federal Aviation Administration (FAA), U.S. Navy, U.S. Air Force, and commercial space industry. After examining issues related to safety, bureaucracy, and ground system reliability and modernization, the task force decided that launch range safety deserved a second look.

OBJECTIVES

The Air Force would like launch range operations to be more efficient and more responsive to commercial and other external users without affecting testing and military launch capabilities. Meeting this goal will require reevaluating the 50-year legacy of the ranges in light of new technologies, lessons learned, and the growing demand for commercial launch services. In support of this effort, the NRC appointed the Committee on Space Launch Range Safety to examine the technologies and procedures used to provide for public safety during space launch operations and to recommend ways to reduce costs and improve efficiency without compromising public safety. The following tasks were assigned to the committee:

1. Conduct a top-level, independent review of the Air Force's safety guidelines and procedures for government and commercial space launches as published in *Eastern and Western Range Safety Requirements* (EWR 127-1, 1997) to determine if there are alternative approaches to the protection of the general public that are both more efficient and less expensive.

2. Conduct an independent assessment of the current and planned range safety and flight termination systems and procedures for government and commercial space launches to estimate the technical feasibility and cost effectiveness of a GPS-based, autonomous flight termination system.

3. Conduct an independent examination of the Air Force's safety guidelines and procedures associated with incursions of aircraft and ships into restricted airspace and waters to determine if holds and delays of government and commercial space launches can be reduced while maintaining an acceptable level of safety.

STUDY PROCESSES AND APPROACH

To execute these tasks, the NRC assembled a panel of 12 experts in launch range safety; space launch operations; launch vehicle systems engineering; launch vehicle guidance, navigation, and control systems; global positioning system (GPS) technology; telemetry, tracking, and command systems technology; risk assessment; and public safety. Some committee members had decades of experience as launch range users or operators, and others had little or no prior experience with space launch systems or operations.

The study benefited from an exceptional amount of public interest and input; more than 100 individuals from interested organizations and members of the general public attended public committee meetings (see Appendix C). The committee welcomed these expressions of interest and invited comments from all meeting participants on the issues under investigation. The committee met four times—in Colorado Springs; Cocoa Beach, Florida; Santa Maria, California (near Vandenberg Air Force Base); and Washington, D.C. The committee received input from range safety personnel on both coasts, commercial space launch providers, range contractors, legal counsel, Air Force Space Command (AFSPC), Air Force Space and Missile Systems Center (SMC), the FAA, the National Aeronautics and Space Administration (NASA), the U.S. Navy Trident Missile Program, and others. In addition, committee members met singly and in small groups with consultants, commercial launch providers, oil rig operators affected by launches from Vandenberg Air Force Base, representatives of Arianespace, range contractors, and the American Institute of Aeronautics and Astronautics. The committee evaluated current range safety systems and alternative approaches with upgraded radars and GPS tracking systems. The committee did not investigate using other types of advanced technologies that may become available in the future.

This report focuses on just one aspect of U.S. space launch capabilities: range safety. The committee's task is not exclusive to this committee. At least 15 recently completed or ongoing studies are also examining national space launch activities (see Appendix D). The Air Force has the difficult task of integrating the analyses, findings, and recommendations of these studies, as well as the perspectives of others interested in the future of the ER and WR.

ORGANIZATION OF THIS REPORT

Through a sometimes lively process of discussion and debate, the committee forged a consensus on each of the findings and recommendations included in this report. Chapter 2 provides background information for readers who are not familiar with launch range operations and safety issues. In response to task 1, Chapter 3 describes risk management approaches to safety. Key elements of this approach include transforming EWR 127-1 into a performance-based requirements document; consolidating within Air Force Materiel Command (AFMC) acquisition-like functions related to range safety, many of which are now being performed by the AFSPC range safety offices at the Eastern Range (ER) and Western Range (WR); and managing risk to meet accepted risk standards rather than to avoid risk whenever possible. Chapter 3 also provides a safety-based rationale for eliminating the expense of downrange assets by moving the Africa gates uprange and advocates the adoption of GPS receiver systems for vehicle tracking. Task 2 is addressed in Chapter 4, which examines the methodology and criteria for flight termination and flight safety systems. Task 3 is covered in Chapter 5, which suggests ways to reduce the impact of intruders on the ranges.

A list of all findings and recommendations appears in Appendix A. Short biographies of committee members are included in Appendix B. Meeting participants are listed in Appendix C. Additional supporting material on related studies and safety models appears in Appendices D and E, respectively.

REFERENCES

EWR 127-1 (Eastern and Western Range Safety Requirements). 1997. Available on line at: *http://www.pafb.af.mil/45sw/rangesafety/ewr97.htm* January 20, 2000.

USAF (U.S. Air Force). 1998. Range Integrated Product Team Report. Peterson Air Force Base, Colo.: Air Force Space Command. Available on line at: *http://www4.nas.edu/cets/asebhome.nsf/web/aseb_related_links?OpenDocument* March 10, 2000.

2

Background

The fundamental goal of U.S. commercial space policy is to support and enhance U.S. economic competitiveness in space activities while protecting U.S. national security and foreign policy interests. . . . Assuring reliable and affordable access to space through U.S. space transportation capabilities is fundamental to achieving national space policy goals. Therefore, the United States will . . . promote reduction in the cost of current space transportation systems while improving their reliability, operability, responsiveness, and safety (NSTC, 1996).

This chapter provides background information for readers who may not be familiar with space launch activities at the ER or WR. Individual sections describe national space launch policy, how responsibilities are divided between the ranges and users, the basis for safety standards, and commercial cost drivers associated with space launch.

NATIONAL SPACE LAUNCH POLICY

In the past, the U.S. space launch industry was dominated by missions sponsored by the Air Force, NASA, and other federal agencies. Now, however, the space launch industry is rapidly becoming a commercial enterprise in which the government emulates commercial customers, shifting from direct management of launch programs to the purchase of space launch services from U.S. commercial launch companies in an open, competitive market. The President's 1994 space policy describes the new scenario:

> U.S. Government agencies, in acquiring space-launch related capabilities, will, to the extent feasible and consistent with mission requirements, involve the private sector in the design and development of space transportation capabilities, encourage private sector financing, . . . [and] encourage private sector and state and local government investment and participation in the development and improvement of U.S. launch systems and infrastructure (NSTC, 1994).

In response to this policy and the underlying economic realities, the three primary space launch customers—the U.S. Department of Defense (DoD), NASA, and the private sector—are moving toward purely commercial modes of operation:

- DoD space-vehicle acquisition programs are increasingly purchasing space launch services instead of launch vehicles. The U.S. Navy has already changed entirely to this mode. According to current plans, DoD will be 100 percent reliant on the commercial space launch industry by 2004 when the last heavy-lift Titan IV has been launched and the new Atlas V and Delta IV launch vehicles are in operation.
- NASA has already shifted entirely to commercial launch vehicles for its unmanned launches, and the space shuttle is transitioning to private-sector operation and maintenance. New reusable launch vehicles (RLVs) are being developed as commercial ventures.
- The private sector market for launching commercial payloads continues to expand as the information age looks for a ride into space. This is a global market in a global economy, and the United States must succeed in commercial terms to maintain a strong space launch position.

Because the ongoing competitiveness of commercial space launch in the United States is important to a broad range of commercial and government activities, the Air Force is committed to improving the cost effectiveness of range operations. However, issues beyond the control of the Air Force limit what the Air Force can accomplish on its own. These issues include:

- aligning national missions to allow the ER and WR to support both commercial and government launches efficiently
- developing a national standard for launch range safety with consistent and universal principles at all U.S. ranges

- increasing the funding priority assigned to range modernization consistent with the importance of maintaining a robust and competitive space launch capability

Issues such as these are being examined by the President's Office of Science and Technology Policy, the FAA, the media, federally funded research and development centers, the American Institute of Aeronautics and Astronautics, Congress, and others (see Appendix D).

Support of Commercial and Government Launches

The Commercial Space Launch Act of 1984, 49 U.S.C. Subtitle IX (as amended), makes the U.S. Department of Transportation responsible for licensing and regulating nongovernment launch activities conducted in the United States (or anywhere in the world if a U.S. corporation controls the launch) and for the reentry of RLVs. The act assigned the FAA the responsibility of issuing safety approvals for launch vehicles, safety systems, processes, services, and personnel. Intended to encourage the commercial space industry and increase access to range facilities, the act specifies that "the Secretary of Transportation shall facilitate and encourage the acquisition by the private sector and state governments of (a) launch property of the U.S. government that is excess or otherwise is not needed for public use; and (b) launch services, including utilities, of the government otherwise not needed for public use." This policy subordinates commercial space launches to government missions at the WR and ER, a limitation that does not reflect the growing importance of commercial space launches. Recognizing the launch ranges as national assets for which the Air Force serves as a steward and rewriting the range mission to put commercial launches on an equal footing with other launches would better align the mission statement with the actual role of the ranges.

National Standards

The convergence of existing range safety standards into a single national standard for commercial and government launches could simplify the safety process faced by users who launch from more than one range. Philosophically, the need for safety is the same whether a vehicle is launched from Florida, Montana, California, Alaska, or anywhere in between. The population density and environmental conditions are different, but the same level of safety should be provided. Commercial companies that use the WR or ER must meet the standards imposed by the Air Force (i.e., EWR 127-1 and related documents). In addition, every contractor has its own safety regulations and must abide by local and federal laws. In many cases contractors are bound by laws from their home states even when using a range in another state. Rules and regulations that are considered acceptable off base at commercial locations should, in general, also be acceptable on base. Fundamental public safety standards should be the same no matter where the operation is conducted.

Several launch locations, such as NASA's Wallops Island, have their own safety documentation, much of it originating from EWR 127-1. Numerous U.S. launch sites, including Wallops Island and new launch sites operated or being considered by new launch service companies, are not under the control of the 30th or 45th Space Wings or any other part of the Air Force. An important issue, then, becomes whether the contents of EWR 127-1 are necessary and sufficient outside Air Force authority. Some documents from other ranges eliminate many of the design solutions included in EWR 127-1 in favor of simply stating performance requirements.

The FAA, which is responsible for licensing commercial launches, has undertaken an initiative to develop commercial launch standards that would apply nationally. The FAA's Office of Commercial Space Transportation has amended its licensing regulation to address commercial launches from federal launch ranges. It also has released notices of proposed rule-making for licensing commercial launch sites and commercial RLVs. The Air Force is helping the FAA develop regulations related to launches from nonfederal launch sites.

Even if a national standard were created, it is not clear how it would be used. A national standard could be referenced by the FAA, but a standard broad enough to be applied nationally would require detailed implementation guidance at specific sites. Also, it is not clear how a national standard would affect the process for updating EWR 127-1, the level of detail used to specify requirements in EWR 127-1, or how those requirements are enforced. To help answer questions such as these, the American Institute of Aeronautics and Astronautics developed an industry consensus on a national standard (see Appendix D).

Priority of Range Modernization

Congress is funding modernization of the ER and WR to reduce recurring costs and to keep the United States globally competitive. Under the Range Standardization and Automation (RSA) modernization program, the Air Force is planning to update satellite data relay systems, improve safety equipment, and reduce turnaround time (the time required to reconfigure ground systems between launches). However, the modernization schedule has been extended several times because of budgetary constraints and the low priority assigned to this effort.

ROLES AND RESPONSIBILITIES OF THE RANGES AND USERS

The ER, with its launch base at Cape Canaveral Air Station, is under the cognizance of the 45th Space Wing, headquartered at Patrick Air Force Base, Florida. The WR's launch base is at Vandenberg Air Force Base, California, and is under the cognizance of the 30th Space Wing. Both wings report to the 14th Air Force and AFSPC.

The history of these two great ranges is the history of the American space program and its contribution to national defense, the ending of the Cold War, and our current space-based military and economic capabilities. A good discussion of the history and current status of the WR and ER can be found in Chapters 4, 5, and 6 of the *Range IPT Report* (USAF, 1998).

Although the ER and WR are considered here primarily in their space launch role, both ranges also function in their historic role as test ranges for national defense systems. The ER is the site for all test launches of Trident submarine launched ballistic missiles (SLBMs) by the United States and the United Kingdom. The WR is the site for all test launches of U.S. intercontinental ballistic missiles (ICBMs). Over the years, both ranges have supported research, development, and training for a myriad of other programs associated with missiles, aircraft, rockets, and other weapons. Within the DoD, the WR and ER are managed as major range and test facility bases under DoD Directive 3200.11, "Use, Management, and Operation of DoD Major Ranges and Test Facilities" (Paragraph 4.2.9.8.). Significantly, as national space ranges, they also come under the purview of DoD Directive 3230.03, "DoD Support for Commercial Space Launch Activities," and the mission charter of AFSPC.

Only recently have the WR and ER been managed and operated using a common range safety document. Their range safety programs did not begin to converge until the late 1980s, and no common document governing range safety at both ranges existed before 1995, when EWR 127-1 was issued.

The 30th and the 45th Space Wings, who manage and operate the ranges, provide the following services for launch customers:

- municipal services and infrastructure necessary to conduct a launch campaign at the site
- management of the siting of the launch pad (explosive arcs, environmental clearances, etc.)
- a permissive environment for the launch entity's acquisition of necessary support services on the ground, either through wing contracts (for government launches using traditional acquisition programs) or launch-customer contracts and purchases (for commercial launches and government launches obtained through commercial launch service contracts)
- range equipment, systems, and services to monitor and track space launches and to ensure public safety during launch; commercial customers reimburse the government for the wing's marginal costs of providing these services for each launch

For all launches, EWR 127-1 vests full authority and responsibility for public safety in the wing commander. The chief of safety at each wing serves as the commander's designated representative to carry out the range safety program, which includes the following tasks:

- enforcing public safety requirements and defining launch area safety and launch-complex requirements for mission flight control and other launch support operations
- reviewing and coordinating changes with range users and providing range safety approvals for operational procedures, as well as oversight of all prelaunch operations at the launch complex and launch vehicle or payload processing facilities as they relate to the safety of the public and launch area.
- reviewing, providing range safety approval, and auditing operations at a launch complex and associated support facilities for launch-complex safety concerns in accordance with launch-complex safety training and certification programs.

The responsibilities of the safety offices at both ranges encompass three functional areas, each of which is involved in protecting the public. *System safety* reviews and approves the design and implementation of safety systems in all launch systems. *Flight safety*, which is responsible for the safety of flight operations, includes planning launch support, establishing allowable mission parameters, monitoring the vehicle and ground systems during countdown and flight, and issuing flight termination commands when necessary. *Ground safety*, which is responsible for the safety of ground operations, includes industrial safety and responding to emergencies during ground operations.

Under EWR 127-1 range users have the following responsibilities:

- providing safe systems, equipment, facilities, and materials
- conducting operations in a manner that is safe and complies with applicable portions of the range safety program
- obtaining reviews and approvals of all safety documents for their programs
- submitting data for flight control operations, obtaining range safety approval, and participating in safety-critical operations
- complying with all other applicable laws and regulations

Commercial range users are highly motivated to carry out these responsibilities in a way that maintains high levels of safety. As noted in the *Range IPT Report*:

> Commercial users have an interest as great or greater than the government in operating launch sites and in conducting launch campaigns safely. A flight failure or safety incident severely impacts the launch manifest and the ability of the

commercial user to attract new or follow-on business (USAF, 1998).

For DoD launches, the 30th and 45th Space Wings have exercised a significant mission-assurance role in the past, but this role has been diminishing recently because of changes in Air Force procurement practices and organization. NASA assumes mission-assurance responsibility for shuttle launches and monitors the work of commercial launch providers for its other launches. For commercial launches, however, the wings have not been assigned any roles or responsibilities associated with mission assurance. The success or failure of commercial launches is the responsibility of the launch company. The mission of the wings is to ensure that every flight can be terminated safely—if it fails. Range safety requirements and responsibilities are related to mission assurance but should be distinguished from nonsafety functions (such as mission assurance) so that safety offices can focus on their primary task (i.e., safety), and range users can be flexible in resolving mission assurance concerns.

SAFETY STANDARDS

Space launch is a potentially dangerous business. Early in the space program risks were largely unknown, and, as a precaution, isolated areas were selected as launch sites. The range safety program has developed to its present state in response to four factors:

- increasing range and explosive power of launch vehicles
- increasing encroachment of civilian populations and municipalities on the launch sites
- increasing sensitivity to public risk
- growing concern that a serious accident involving the general public would inhibit important space programs

An early goal used to define launch safety standards was to ensure that the public would be subject to no more risk from space launches "than that imposed by the overflight of conventional aircraft" (EWR 127-1, 1997). This goal has since been broadened to ensuring that the risk to the general public from space launches be no higher "than the risk voluntarily accepted in normal day-to-day activities" (EWR 127-1, 1997). The foundation for both goals is rooted in the legislative history of Public Law 60 from the 81st Congress, but neither goal is legally mandated. Even so, the ranges have well-defined safety standards based on evolving analysis techniques, technology, and regulations. Two standards at the heart of the current range safety programs at the two ranges are critical to the discussions in this report:

- The standard of collective risk to the general public, expressed in terms of *casualty expectation* (E_c), must be less than 30×10^{-6} for each launch. This implies that one serious injury or fatality can be expected to occur for every 33,000 launches, or once every 1,000 years for a launch rate of 33 per year.[1]
- The overall goal for reliability of the flight termination system (FTS) is 0.9981, with air and ground subsystems each meeting a reliability of 0.999.[2]

Range safety is based on two primary, complementary elements:

- The launch hardware is designed to be safe and is then managed, maintained, and operated to ensure that the design levels of safety are achieved at all times.
- The range operators can reliably monitor launches in progress, and, if something happens in flight that could compromise public safety, they can shut down the vehicle propulsion system so the vehicle follows a ballistic flight path to a known, safe impact point.

Finding 2-1. Range safety personnel and procedures have well protected people and property. In the history of the U.S. space program, no members of the general public or launch site workers have been killed or seriously injured during a launch accident.

COMMERCIAL COST DRIVERS

Cost will be a key driver in future space launch competitions, and range costs are an important element of total launch costs. Range operators and commercial launch customers incur three kinds of launch costs:

- The cost of maintaining and operating the ER and WR, which during fiscal year 1998 was $731 million, exclusive of military pay. More than half of the budget ($418 million) was dedicated to infrastructure. The remaining $313 million was spent on range and launch support contracts. Of the total, $603 million was an uncompensated budget burden on the Air Force and U.S. taxpayers (USAF, 1998). The Air Force's desire to reduce this burden through more efficient and less costly range safety processes is one of the factors behind current efforts, including this study, to improve range efficiency.
- Reimbursable costs, which are paid to AFSPC by range users to offset the cost of range operations. This amount was $128 million in fiscal year 1998 (USAF, 1998). The amount paid by commercial launchers is a business burden directly related to global competitiveness.
- The launch companies' internal costs of complying with range safety rules, which are another direct business burden.

---

[1]Collective and individual risk are discussed in more detail in Chapter 3.
[2]FTS reliability is discussed in more detail in Chapter 4.

Launch companies also face the opportunity cost of business lost to foreign competitors with less costly range safety systems and processes. An important by-product of establishing a more efficient range safety system would be improved U.S. competitiveness.

Inconsistencies between the two ranges and between the range safety and acquisition organizations result in redundant and sometimes conflicting or risk-averse requirements being levied on the launch providers, which increases range certification costs and operational complexity.

Range safety requirements impact user costs in a number of ways. A user may wish to launch a new vehicle or a derivative of an existing vehicle on one of the ranges. The Atlas V is an example of a derivative vehicle currently under development. The total cost for range safety certification is estimated by Lockheed Martin at $1.8 million, of which the flight segment cost will be approximately $600,000 (not including launch operations procedures). Lockheed Martin has estimated that there are more than 18,000 requirements in EWR 127-1, of which 10,778 had to be addressed individually for the Atlas V.

A change in one segment of a launch vehicle can require recertification of all vehicle range safety systems to the latest requirements, not the requirements that were current when those systems were originally certified. For example, the booster for the Taurus launch vehicle, which was certified for flight on the WR, was upgraded from a Minuteman to a Castor 120 stage. Recertification for both ranges required an upgrade from 1989 certification standards to the 1995 version of EWR 127-1. Total nonrecurring costs for the recertification were more than $1 million, including more than $750,000 changes to onboard hardware. These changes also increased recurring costs by $60,000 per vehicle.

A launch vehicle certified for flight on one range still requires separate certification for flight on the other range. The process of certifying the Atlas II for flight on the WR began in 1993, after it had been certified and flown successfully from the ER. The certification process still was not complete in mid-1999. At that time, the cost associated with recertifying the Atlas II for the WR had exceeded $1 million. Eighty percent of that cost was related to meeting the flight safety requirements in Chapter 4 of EWR 127-1.

For the Atlas IIAS, the WR required recertification of the FTS antennas, even though antennas of that design had been flying for nearly 30 years on both coasts. Lockheed Martin reported that the WR required qualification testing of three sets of hardware at a total cost of $300,000. (Unexpected problems unrelated to antenna performance were encountered during testing, which added to the cost.) Ultimately, the antennas were certified with no changes in design (Smith, 1999). In addition, the WR required recertification of the Atlas IIAS electrical box shock mount isolators, which had flown successfully at the ER, for use with FTS systems. Testing of the mounts and boxes cost an additional $300,000. Going the other way, the manufacturer of the Pegasus launch vehicle estimated that certifying the Pegasus for launch at the ER cost about $1 million and took more than six months and four labor-years of engineering staff time, even though the Pegasus had been launched many times from the WR.

Range safety offices may also require that users pay for tests to assess the effectiveness of specific design solutions. The WR requires that ground box acceptance testing of each FTS component be conducted in a special facility provided by the WR. This facility also is used for retesting to extend the shelf life of range safety components when required to meet launch schedules. At the ER, however, users have the option of conducting these tests as part of their normal vehicle manufacturing process. Also, the WR process requires that users provide the range with a new test set or modify an existing test set whenever changes are made to the design of the receiver or FTS. In one example cited by Lockheed Martin, the costs of providing a test set to the WR exceeded $1 million. On the positive side, Orbital uses the WR to test all Pegasus FTS boxes regardless of the ultimate launch site, obviating the need for duplicate test facilities and equipment but entailing additional transportation time and expense for launches conducted anywhere except the WR.

Many factors affect the cost and effort of certifying a launch vehicle on one range after it has operated from the other range. The committee did not conduct an independent analysis of the need for the testing described above and does not assert that any or all of it was arbitrary or capricious. Clearly, however, there is room for improvement in developing common test and certification practices, and more standardized requirements between the ranges would help address this problem.

REFERENCES

DoD (U.S. Department of Defense). 1998. Use, Management, and Operation of DoD Major Ranges and Test Facilities. Department of Defense Directive 3200.11. January 26, 1998. Washington, D.C.: U.S. Department of Defense.

DoD. 1986. Department of Defense Support for Commercial Space Launch Activities. Department of Defense Directive 3230.03, October 14, 1986. Washington, D.C.: U.S. Department of Defense.

EWR 127-1 (Eastern and Western Range Safety Requirements). 1997. Available on line at: *http://www.pafb.af.mil/45sw/rangesafety/ewr97.htm* January 20, 2000.

NSTC (National Science and Technology Council). 1994. National Space Transportation Policy (NSTP-4). August 5, 1994. Washington, D.C.: The White House.

NSTC. 1996. Fact Sheet: National Space Policy. September 19, 1996. Washington, D.C.: The White House.

Smith, D. 1999. Atlas FTS Antenna Requalification. Briefing by Dan Smith, Lockheed Martin Astronautics, to a panel of the Committee on Space Launch Range Safety, Lockheed Martin Astronautics, Denver, Colorado, June 24, 1999.

USAF (U.S. Air Force). 1998. Range Integrated Product Team Report. Peterson Air Force Base, Colo.: Air Force Space Command. Available on line at: *http://www4.nas.edu/cets/asebhome.nsf/web/aseb_related_links?OpenDocument* March 10, 2000.

3

Risk Management Approaches to Safety

All ventures entail some risk. With space launches, this risk applies to the loss of the mission, property damage, or casualties for mission personnel or the public at large. A strict *risk avoidance* stance—reducing risk to the lowest possible level regardless of cost—would preclude space launch by making it unaffordable. *Risk management*, however, is designed to meet standards of acceptable risk based on overall costs and benefits. Risk standards then can be used to derive safety requirements, and old requirements not needed to satisfy risk standards can be eliminated.

This chapter describes a risk management approach to space launch range safety. The starting point is EWR 127-1, the primary range safety document at the ER and WR. Although risk standards specified in EWR 127-1 are consistent with a risk management approach to safety, many of the specific guidelines in EWR 127-1 apply the standards in a way that avoids risk. Shifting the operational implementation of EWR 127-1 from risk avoidance to risk management requires a cultural change.

The focus of the chapter then shifts to the division of roles and responsibilities for range safety between AFSPC and AFMC. The last part of this chapter examines risk criteria, risk management, and analysis methods, including the potential for eliminating downrange safety-related assets at the ER.

PHILOSOPHY OF EWR 127-1

EWR 127-1, which was most recently updated on October 31, 1997, is based on earlier range safety manuals developed independently at the ER and WR. Rather than requiring that each user develop its own methods of compliance, the ranges defined many design solutions and included them in EWR 127-1 as requirements. Also, to reduce the need to refer to regulations and instructions issued by other government organizations, many of their requirements were quoted, expanded, or paraphrased and inserted into EWR 127-1. As a result, EWR 127-1 is a huge document that is focused much more on methods and solutions than on basic, performance-based safety requirements. Also, EWR 127-1 has two sets of requirements in many areas, one for the WR and another for the ER.

EWR 127-1 is issued under the authority of the 30th and 45th wing commanders. The Air Force plans to issue the next revision of EWR 127-1 under the authority of AFSPC, which also will be involved in developing future updates. The committee supports this plan, which should enhance the ongoing convergence of documents issued by and requirements established by the WR and ER.

The fundamental safety standard in EWR 127-1 is the collective risk criteria, E_c, of 30×10^{-6}. However, the safety philosophy and practices codified in EWR 127-1 often go beyond what is necessary to meet that standard. For example, risk management becomes risk avoidance when EWR 127-1 speaks of "risk minimized to the greatest extent possible." Also, Chapter 4 of EWR 127-1 laboriously lays out hardware, construction, and test requirements for vehicle safety systems. Detailed, often step-by-step procedures and processes are dictated in annexes. Although EWR 127-1 is based on limiting collective risk to the general public, E_c, to less than 30×10^{-6} for each launch, no allowable component- or system-level risk assessment is provided, and the "highest achievable system reliability" has become the *de facto* guiding principle.

EWR 127-1 does not describe the source of most of its requirements. In parallel with this study, SMC, which is part of AFMC, initiated a study to document the sources of requirements, determine which requirements are design solutions, and identify the actual standards represented by design solutions. The committee endorses this and other efforts to determine the validity of specific requirements in light of industry standards and existing laws and regulations.

Another complicating factor is the process of "tailoring," which allows alternate means of complying with the requirements of EWR 127-1. The tailoring process has evolved to the point that, in essence, a unique version of EWR 127-1 is created for each new launch system. Tailoring provides range users with great flexibility, but it also reveals a serious shortcoming in the usability of EWR 127-1: range safety requirements are defined on an *ad hoc* basis by the safety offices (during the tailoring process) rather than in published regulations.

In addition, inconsistencies in the tailoring process may mean that different users incur different costs to certify the same equipment depending, in part, on the negotiating skills and expertise of the engineers working with range safety personnel. A C-band radar beacon may cost $60,000 on one certified vehicle and less than $20,000 on another because different requirements for parts quality and acceptance testing are established during the tailoring process. Users may continue to use more expensive components because the cost of certifying lower cost components and recertifying other related hardware on the vehicle would wipe out the savings of using the lower cost components.

Reformulating EWR 127-1 as a performance-based requirements document would have several benefits. The need for tailoring, as it is currently practiced, could be greatly reduced or eliminated. The number of individual requirements in EWR 127-1, which add to the costs borne by both the Air Force and the launch customer, would be greatly reduced. A clear distinction would be made between non-negotiable performance-based requirements and approved methods of compliance that can be waived if an equally effective alternative is available. Users would have the option of (1) implementing the approved method of compliance to streamline the review process, or (2) using an alternate means of compliance, for which users would accept the responsibility for getting approval.

Primary Recommendation on EWR 127-1. AFSPC should simplify EWR 127-1 so that all requirements are performance based and consistent with both established risk standards for space launch (e.g., E_c of 30×10^{-6}) and objective industry standards. The process of revising EWR 127-1 should include the following steps:

- Eliminate requirements that cannot be validated.
- Remove all design solutions from EWR 127-1.
- Establish a range user's handbook or other controlled document to capture lessons learned and design solutions recognized by the ranges as acceptable means of compliance. (Requirements should be retained in EWR 127-1.)
- Form a joint government/industry team to establish procedures for periodically updating EWR 127-1 and ensuring that future requirements are performance based.
- Converge the modeling and analysis approaches, tools, assumptions, and operational procedures used at the Western and Eastern Ranges.

ROLES AND RESPONSIBILITIES OF THE AIR FORCE SPACE COMMAND AND AIR FORCE MATERIEL COMMAND

EWR 127-1 includes a great deal of detailed information on organizational roles and responsibilities. Briefings and documents provided to the committee also included significant amounts of information on these subjects, and the committee reviewed two versions of an AFSPC/AFMC memorandum of agreement (MOA) to understand the division of spacelift roles and responsibilities. One of the agreements was signed in 1997 and is currently in effect (AFSPC/AFMC, 1997). The second was an unsigned draft of an updated MOA dated May 1999 (AFSPC/AFMC, 1999).[1] The purpose of the review was to determine if changes in roles and responsibilities might improve the efficiency of range safety operations. The review focused on two alternatives: maintaining the status quo and adopting the approach defined in the MOAs.

Both MOAs confirm the intent that AFMC assume responsibility for the acquisition, developmental testing, sustainment, and improvement of launch vehicles, spacecraft, and launch range systems. The process of transferring related functions from AFSPC to AFMC began in earnest with the range modernization program. Nonetheless, developmental engineering continues to be performed by AFSPC in the area of safety systems for launch vehicles and spacecraft. AFSPC has performed these tasks since it was established more than a decade ago, even though they overlap existing AFMC functional responsibilities.

The transfer of the development and engineering functions related to range safety from AFSPC to AFMC would be consistent with the intent of the MOAs and with normal Air Force practices. If properly executed, this transfer would reduce costs and workload for both the Air Force and range users by eliminating duplicative efforts and standardizing procedures and systems.

Air Force Memorandum of Agreement on Spacelift

The MOA on spacelift between AFSPC and AFMC delineates in great detail the roles and responsibilities of each organization. The basic concept of operation states that AFMC will develop and acquire space systems based on approved AFSPC requirements and that AFSPC will conduct spacelift operations to meet its war-fighting requirements. Each command is expected to manage the hardware, software, and support necessary to meet its mission requirements. AFMC's role is to "perform all functions required to acquire, conduct developmental testing, sustain, and improve the operational performance of launch vehicles, satellites, and launch range systems" (AFSPC/AFMC, 1997). AFMC

---

[1] Instead of issuing a new MOA based on the 1999 revision, the Air Force may decide to issue an Air Force Instruction.

is also responsible for contracting for these items as required and for the full life cycle of the acquisition process with support from AFSPC. AFMC is responsible for development, qualification, and acceptance tests of new or modified systems to show that systems comply with all specifications and requirements provided by the operational organization at the component, subsystem, and system levels. AFMC is also responsible for the provision of sustaining engineering and depot-level maintenance for launch vehicles, upper stages, range systems, and associated ground equipment.

In support of the acquisition process, AFSPC is responsible for defining and prioritizing operational requirements for existing and new launch vehicles and systems and for communicating those requirements to AFMC. AFSPC is responsible for developing, planning, and conducting operational testing and evaluation to demonstrate systems' operational effectiveness and suitability under realistic conditions. AFSPC's primary role (in terms of space launch) is to conduct prelaunch and launch operations with AFMC support as necessary to resolve anomalies.

It is noteworthy that the 1997 MOA does not delegate to AFSPC a special role in the development of safety systems. The MOA makes AFMC responsible for providing a complete launch vehicle, including safety systems, that meet AFSPC requirements. AFMC is the office of primary responsibility until "space system asset availability," which is the program milestone when hardware is turned over to AFSPC. At that time, the 1997 MOA specifies that hardware should be ready for flight except for prelaunch processing. The May 1999 draft of the new MOA does not indicate any changes in the division of responsibilities described above (see Figure 3-1).

Both MOAs are also consistent with the normal roles and responsibilities assigned to a system program director in AFMC and other DoD acquisition commands. System program directors are responsible for all aspects of new system

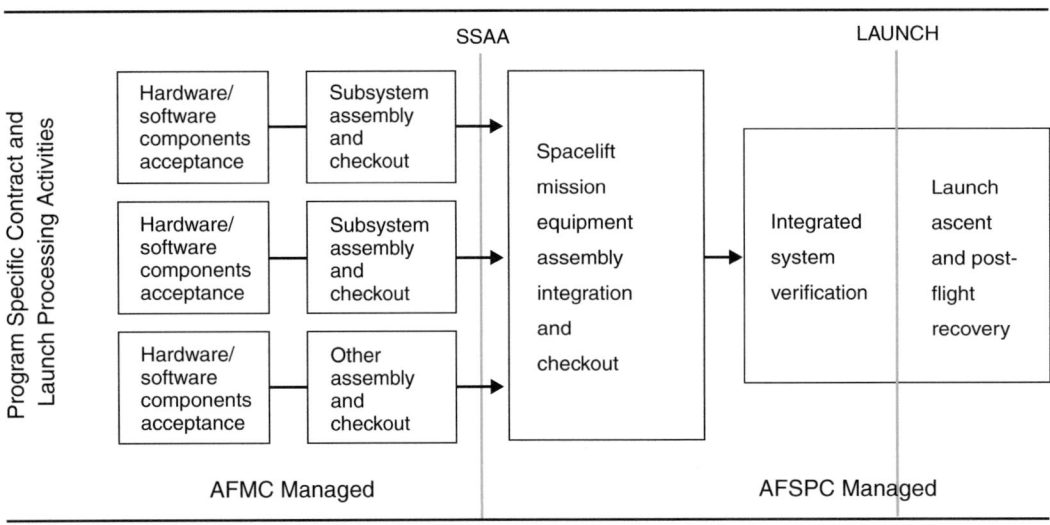

FIGURE 3-1 Air Force roles and responsibilities for space launch. Source: AFSPC/AFMC, 1999.

acquisition, including developmental testing, sustainment, and overall cost effectiveness. Before systems are turned over to operational commands, AFMC must certify that all systems have been designed to meet operational requirements and check out at the component, subsystem, and system levels. System program directors' responsibilities extend to all subsystems and include safety, including safety of flight where applicable (AFMC, 1998).

Development and Engineering Functions at the Ranges

The ranges, which have a long history of development and developmental testing, for many years were part of an Air Force acquisition command. Over the years, research and development related to ICBMs, SLBMs, and other space launch systems have been reduced, and operations have became increasingly important. In 1982 a space operations command (i.e., AFSPC) was created, and in 1990 it assumed command of the ranges. Soon thereafter responsibilities for development and sustaining engineering of range instrumentation were transferred back to an acquisition command (i.e., AFMC), which is managing the ongoing range modernization program. In accordance with the 1997 MOA, AFSPC sets programmatic requirements for the modernization effort and supports the acquisition.

In contrast, personnel from launch vehicle manufacturers and the ranges indicated to the committee that the safety offices at both ranges (which are part of AFSPC) have assumed essentially full responsibility for analysis and testing of safety systems to certify compliance with requirements in EWR 127-1. Areas of particular interest include onboard safety systems, such as FTSs, receivers, batteries, and tracking devices. As described earlier in this chapter, obtaining authorization from either range safety office to use new or modified systems can be labor-intensive for both users and the range safety offices, can significantly increase user costs, and can take months or years to complete, even if the "new" system has been previously authorized for use at the other range. The range's heavy involvement in analysis, testing, and certification results in duplication of effort, because AFMC and the individual system program directors already have responsibility for approving flight safety systems along with other vehicle subsystems. Current practices are also at odds with normal Air Force acquisition practices and with specific guidance in both versions of the MOA on spacelift roles and responsibilities.

Transfer of Acquisition Functions to an Acquisition Command

Analyzing, testing, and certifying the design of new and modified systems involves staff assigned to many different elements of the range safety offices at the WR and ER. The committee recommends that the Air Force transfer acquisition-like functions related to range safety from AFSPC (i.e., from the ranges) to AFMC. (See Primary Recommendation on Roles and Responsibilities, below.) The committee recognizes that determining which functions are involved in the development phase and where to draw the line organizationally between operations and development will be difficult because these functions have been closely linked for many years. The Air Force will have to decide where in AFMC range safety functions should reside and what changes should be made in the size of the workforces and budgets of AFMC and AFSPC. It will be important to establish a concept of operations that ensures safety decisions are objective and provides for effective communications between acquisition and operational commands involved in range safety. Although quick action is needed, the committee believes that the Air Force should carefully review implementation issues such as these before moving forward with the recommended transfer of responsibilities. Also, it should be emphasized that the recommended transfer is about functions—not existing organizations or individuals.

In addition to a transfer of safety system functions, the committee concluded that responsibilities for developing detailed safety models to support flight operations should also be transferred. The range safety staff must, of course, be very knowledgeable about the content of the models and how to use them during prelaunch and launch operations. Also, close working relationships between operational staff (at the ranges) and acquisition staff (within AFMC) must be maintained to ensure that new systems and system modifications are consistent with operational needs and can be efficiently implemented in an operational setting.

The basic responsibility of the range safety offices—the protection of human life and property—should not be changed. The basic responsibility of the newly established AFMC safety office would be to support to the system program offices (and AFSPC, as appropriate) during the acquisition, developmental testing, sustainment, and improvement of space systems. The AFMC safety office would also be responsible for certifying safety readiness for other government, civil, and commercial launch operations at the WR and ER, as outlined below. Requirements for design, qualification, and acceptance testing processes would be removed from EWR 127-1 and documented in an AFMC handbook describing acceptable means of compliance and lessons learned, in a manner consistent with the Primary Recommendation on EWR 127-1. AFMC would certify to AFSPC that new and modified systems have met specified requirements. Systems would then be handed over to AFSPC for prelaunch and launch operations, including operational testing, in accordance with the current MOA.

Existing and planned independent functional organizations in AFMC and SMC could be used as a model for establishing a safety office. To ensure that safety decisions are objective, the safety office must be independent. This could be achieved by allowing safety managers to report unresolved safety concerns directly to a high level within the chain of command without putting their jobs on the line. In addition to supporting system program offices and the ranges, the safety office should be responsible for centralizing and simplifying the development of safety policy, procedures, and systems; maintaining a strong engineering, analysis, modeling, and simulation staff through training and career advancement; and reducing costs for the range operators and users.

Other Users

The discussion above is focused on satisfying the safety needs of space launches by the Air Force on the WR and ER. However, the AFMC safety office described above would also have to certify the safety of flight and ground systems for commercial space launches and other activities at the ER and WR (e.g., launches of the NASA spacecraft, ballistic missile tests by the U.S. Navy and Air Force, and aircraft flight tests). Certification of these systems would be based on FAA space launch regulations (for commercial launches),[2] EWR 127-1, other pertinent documents, and the results of design reviews, analyses, developmental tests, and/or operational performance records. Close coordination between operational staffs at the ranges, system operators or developers, and the AFMC safety office would be necessary to evaluate risks, generate new safety tools, establish appropriate risk standards, and manage risk for these missions. In addition, the safety office could also develop, procure, and certify standard flight safety systems and make them available to users.

Working with other involved parties, the Air Force should prepare an instruction or other appropriate document describing the safety group's role in the development and certification of Air Force, government, and commercial safety systems and interfaces with other organizations in AFMC, AFSPC, FAA, NASA, and industry. The specification of design, qualification, and acceptance testing processes, as well as lessons learned, should be incorporated in the new documentation and removed from EWR 127-1.

Findings and Recommendations on Roles and Responsibilities

Finding 3-1. AFSPC has transferred responsibility to AFMC for development, developmental testing and evaluation, and sustaining engineering of range safety ground systems. Organizational responsibilities for many other range safety processes and procedures, however, are inconsistent with the current memorandum of agreement between AFSPC and AFMC on spacelift roles and responsibilities. In addition to the operational workforce, each AFSPC range safety office also has an engineering workforce that establishes flight safety system design and testing requirements and certifies that flight safety systems meet safety requirements at the component, subsystem, and system levels. These acquisition-like functions overlap the responsibilities of AFMC.

Finding 3-2. The complete transfer of range safety development, developmental testing and evaluation, and sustaining engineering to AFMC would, if properly implemented, increase efficiency and reduce costs without compromising safety by eliminating overlapping responsibilities between the ranges and AFMC, by minimizing differences in range safety policies and procedures applicable to the Western and Eastern Ranges, and by enabling users to deal with a single office when seeking approval to use new or modified systems on both ranges.

Primary Recommendation on Roles and Responsibilities. The Air Force should fully implement the memorandum of agreement between AFSPC and AFMC on spacelift roles and responsibilities. This would consolidate in AFMC the acquisition-like functions related to safety that are now performed by AFSPC organizations at the Eastern and Western Ranges. These functions include developmental testing and evaluation, sustaining engineering, and certifying that system designs meet safety requirements. To manage the safety aspects of the acquisition-like functions specified in the memorandum of agreement, AFMC should establish an independent safety office. Operational responsibilities, such as generating safety requirements, operational testing and evaluation, and all prelaunch and launch safety operational functions, would be retained by AFSPC.

Recommendation 3-1. AFSPC should issue an Air Force Instruction addressing the certification of flight safety systems for commercial, civil, and military launches at the Western or Eastern Range. The instruction should include a description of interfaces among responsible organizations, such as AFSPC, AFMC, FAA, NASA, and commercial contractors.

RISK CRITERIA, RISK MANAGEMENT, AND ANALYSIS METHODS

This section discusses several key issues affecting public safety during launch. First, the current risk criteria used by the Air Force are discussed. Next, certain inconsistencies between these accepted risk-management criteria and operational methods based on risk avoidance are described. These inconsistencies are examined in light of the risk posed by vehicles as they approach orbit to show that downrange safety-related assets can be eliminated while safety is maintained within accepted limits. Finally, general safety assessment and modeling issues are presented, followed by an outline of the major differences in modeling and analysis methods at WR and ER.

---

[2]Information on the FAA licensing process, including relevant statutes, regulations, and policies, is available on line (*http://ast.faa.gov/licensing/*).

TABLE 3-1 Comparison of Maximum Acceptable Collective Risks

Activity	Annualized Collective Risk (number of expected fatalities per year)
Space launch (ER and WR)	1×10^{-3}
Commercial nuclear power plants (United States)	2×10^{-6}
Hazardous material storage (Hong Kong)	7×10^{-3}
Nuclear and chemical industry (Netherlands)	1.1×10^{-3}
British Ministry of Defense	6×10^{-3}
Petrochemicals (Santa Barbara County)	1×10^{-3}

Source: RCC, 1997b.

In the course of its study of the risk criteria used at the WR and ER, the committee reviewed a number of documents. Central among them was Chapter 3, "Risk Criteria Rationale," in the Supplement to Range Commanders Council (RCC) Standard 321-97 (RCC, 1997b).[3] This document contains an extensive treatment of the principles and logic behind the use of common risk criteria.

Risk Criteria

According to Air Force Instruction 91-202, "risk should be quantified and acceptable limits established" (USAF, 1991). EWR 127-1 describes the principal risk criterion for space launches at the WR and ER: a casualty expectation, E_c, "of 30×10^{-6} shall be used by both ranges as a level defining 'acceptable launch risk without high management (Range Commander) review.' Based on national need and the approval of the Range Commander/Wing Commanders, launches may be permitted using a predicted risk above 30×10^{-6}" (Paragraph 1.4.1). Previous versions of EWR 127-1 indicated that the upper limit of risk that a commander might approve locally was an E_c of 300×10^{-6}.

The ranges also use an individual risk criteria, P_c, to describe the probability of an individual in any particular place being killed or severely injured during a launch. P_c can be used to determine whether specific personnel are at high risk in a given area. EWR 127-1 prohibits exposing members of the general public to a P_c greater than 1×10^{-6}; the limit for mission-essential personnel is 1×10^{-5}. (See Chapter 5 for a discussion of limits on individual hit probabilities, P_i, for ships and aircraft.)

These collective and individual risk criteria are consistent with RCC Standard 321-97, which recommends their use on all DoD ranges (RCC, 1997a). The supplement to RCC Standard 321-97 also describes acceptable levels of risk in other domains. The supplement cites regulatory procedures promulgated by the U.S. Department of Labor, Environmental Protection Agency, Occupational Safety and Health Administration, and the Food and Drug Administration that pertain to individual and collective risks of industrial, occupational, public, and in-the-home accidents, as well as risk levels related to carcinogens (RCC, 1997b).

The RCC analysis uses annualized risks when comparing space launch range safety to safety in other fields. An E_c of 30×10^{-6} is equivalent to a rate of one casualty every 1,000 years, or 1×10^{-3} casualties per year, given an average launch rate of 33 per year. Table 3-1 compares this risk level with the annualized collective risk limits for other fields and shows that annualized risks on the order of 10^{-3} are commonly accepted, both in the United States and internationally. The significantly lower risk standard established by the Nuclear Regulatory Commission for the operation of nuclear power plants reflects concerns about a major catastrophe that could affect tens or hundreds of thousands of people near a nuclear power plant and the potential long-term consequences of a nuclear accident.

An E_c of 30×10^{-6} is also comparable to the risk accepted by the public for commercial air travel. From 1982 through 1998, U.S. air carriers had 131 million departures, and accidents resulted in 2,868 casualties (354 serious injuries and 2,514 fatalities), which is equivalent to an E_c of 22×10^{-6} per departure (NTSB, 2000, Tables 3 and 5).

Finding 3-3. A collective risk standard (i.e., a casualty expectation, or E_c) of 30×10^{-6} per launch for members of the general public is consistent with the risk standards of many other fields in which the public is involuntarily exposed to risk, both domestically and internationally.

Application of Risk Management

To ensure safety, range safety tracks each launch vehicle and predicts its instantaneous impact point (IIP), which is a real-time estimate of where the vehicle would land if the flight were terminated. Because of the high speed of the

---

[3]The membership of the Range Commanders Council includes the commanders of the ER, the WR, and 19 other test, training, and operational ranges operated by the DoD.

vehicle, the IIP may be several thousand miles downrange of the vehicle's current position. The nominal flight path, the actual course of the vehicle, and the computed IIP change during flight. If for any reason the range safety personnel cannot verify that a vehicle is and will remain within specified boundaries (e.g., if tracking systems fail), they will terminate the flight.

Background on Destruct Lines and the Africa Gates

A key element of current range safety procedures involves defining the thresholds used during launch to determine when a flight should be terminated. These thresholds are ultimately based on impact limit lines (ILLs), which extend downrange from the launch site and define the area in which debris (from planned stage drops, vehicle explosions, or thrust termination) may land. Flight trajectories and ILLs are calculated and approved prior to launch to protect people and property. The ILLs are not explicitly defined by safety metrics (such as E_c). Instead, they are based on risk avoidance: "Whenever possible, the overflight of any inhabited landmasses is discouraged and is approved only if operational requirements make overflight necessary, and risk studies indicate probability of impact and casualty expectancy are acceptable" (EWR 127-1, paragraph 2.3.6). To account for delays in operator response, uncertainties about vehicle breakup, winds, and other aerodynamic effects, destruct lines are defined inside of the ILLs. If a vehicle's IIP reaches a destruct line, the flight is terminated.

Wherever the vehicle passes over inhabited landmasses before orbital insertion, "gates" (i.e., exits) in the ILLs and destruct lines are defined. The vehicle must pass through the gate or the flight will be terminated. The gates are perpendicular to the nominal trajectory, and the width of the gates accounts for tracking uncertainties and acceptable variations in trajectory. The use of gates and their locations are defined by EWR 127-1 and related Air Force documents—the committee knows of no international agreements that require their use.

Even though orbital insertion typically occurs over the horizon, gates and downrange tracking, telemetry, and FTS capability are not needed to satisfy the E_c risk standard (30×10^{-6}) for orbital launches from the WR. Allowable azimuths are constrained during the uprange portion of flight to avoid overflight of inhabited landmasses during the boost phase. Orbital vehicles are tracked and FTS commands are issued only to the horizon (i.e., only as long as uprange systems have direct contact with the vehicle).

DoD Directive 3200.11 requires ranges to prevent launch vehicles from "violating established limits through impact for vehicles with suborbital trajectories and through orbital insertion or escape velocity for space vehicles" (Paragraph 4.2.9.8). The 45th Space Wing made the following statement in response to a query from the committee:

Our interpretation is that DoD Directive 3200.11 drives a definite requirement for downrange assets at the ER to support command destruct . . . and metric tracking . . . . Even interpreting 3200.11 liberally and employing risk management techniques, metric tracking would be required to support notification in cases of accident or errant trajectories. The responsibility for safety from launch to orbital insertion (for space vehicles) and from launch to impact (for ballistic vehicles) is consistent with knowing the vehicle's position and its predicted impact point at all times during these periods of flight. This information would also be necessary for the settlement of international claims or disputes in the event that a malfunction occurs beyond the destruct capabilities of the ranges (45th SW, 1999).

Neither DoD nor AFSPC instructions establish different risk standards for citizens of the United States and citizens of foreign nations, and the ER allows vehicles to proceed over Europe and Africa without further intervention if the vehicles have successfully navigated the appropriate gates. The location of the Africa gate typically corresponds to the position of the IIP at approximately 500 to 700 seconds after launch (see Figure 3-2). Depending upon the launch vehicle and flight azimuth, Africa gates may be as far downrange as 10° west longitude. Downrange radar assets at Antigua and Ascension Islands are required to provide the vehicle position data used to compute the IIP beyond approximately 480 seconds (Figure 3-3). Maintaining, staffing, and operating downrange facilities and providing reliable, real-time communications between the downrange facilities and the Range Operations Control Center (ROCC) at the ER is expensive. Coordinating launch operations with remote facilities also complicates range safety operations and increases the risk of holds and delays (if problems occur at the remote facilities or in the communications links).

Moving the Africa Gates

Several factors suggest that the collective risk standard, E_c, could still be met if the Africa gates were moved uprange. First, most major vehicle events (staging and engine starts) occur within approximately 300 seconds of launch while the vehicle is well within the area covered by uprange assets. Following these events, vehicles have historically been quite reliable. After 300 seconds, for example, the probability of an Atlas failure is estimated at 25.7×10^{-6} per second until the end of the vehicle's mission at 670 sec (see Table 3-2 and Figure 3-4).

Vehicles that successfully complete uprange staging events are highly reliable, and their IIPs are travelling very fast—much faster even than the vehicles themselves as they approach orbital velocity; the IIP disappears as soon as the vehicle reaches orbital velocity. The IIPs of Atlas vehicles, for example, are over Africa for only 0.3 to 8.08 seconds before the vehicle reaches orbital velocity or the IIP enters the Indian Ocean. Based on the vehicle

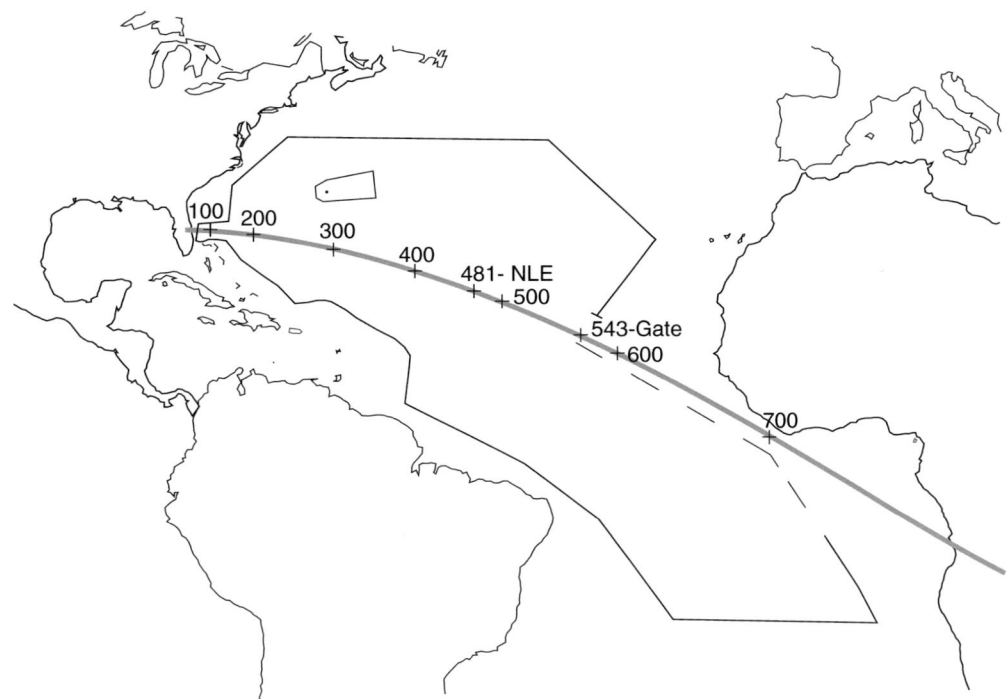

Numbers indicate time after launch in seconds. The IIP encounters the Africa gate 543 seconds after launch. NLE = no longer endanger (if a vehicle somehow reversed course at this time, it would not have enough propulsive energy to return to the continent from which it was launched).

FIGURE 3-2 Instantaneous impact point trace and Africa gate location for Titan IV-B25. Source: 45th SW, 1999.

reliability data in Table 3-2, the probability of debris from an Atlas hitting Africa is less than 210×10^{-6} per launch.[4] This is equivalent to one event in more than 4,800 launches and is consistent with the probability of land impact for Titan launches on similar trajectories, which has been estimated to be approximately 300×10^{-6} per launch after 400 seconds. When combined with the subsequent probabilities of impacting a populated area and causing casualties,[5] the risks from flying over Africa appear to be well within the standard acceptable for the U.S. population, 30×10^{-6} (Ward, 1997). In fact, for an Atlas II/IIAS launch vehicle that successfully reaches the existing Africa gates, E_c for the remainder of flight is 8×10^{-8} (LMA, 1999), and the remaining E_c for an Atlas IIIB when uprange facilities lose contact with the launch vehicle is 4.9×10^{-8}.

Even if a failure were to occur more than 400 seconds after launch, the vehicle is travelling very fast and it would break up from dynamic forces upon reentering the atmosphere. At this stage of flight, fuel cutoff often is used for flight termination instead of explosive charges. Cutting off fuel helps prevent the vehicle from veering off course and minimizes the size of the debris pattern by keeping the vehicle largely intact until it breaks up at lower altitudes. For failure modes in which thrust ends prematurely, a thrust-termination type of FTS would have no added benefit. Therefore, the absence of FTS capability beyond the coverage area of uprange assets would not reduce safety for malfunctions that terminate thrust prematurely. This would not be true if a malfunction occurred downrange that unexpectedly reduced vehicle thrust or directed a vehicle off the intended trajectory while maintaining stable, powered flight. The committee concludes, however, that the vehicle's design characteristics and its high speed at this point in the flight make it highly unlikely that a significant change in IIP would occur before the vehicle breaks up even without intervention by an FTS. This conclusion is supported by calculations of E_c during the downrange portion of flight, as noted above.

The current placement of Africa gates derived from ILLs and destruct lines is based on risk avoidance. From a

---

[4]Calculated as follows: $1-(1-0.0000257)^{8.08}$.
[5]Population modeling is described in Appendix E.

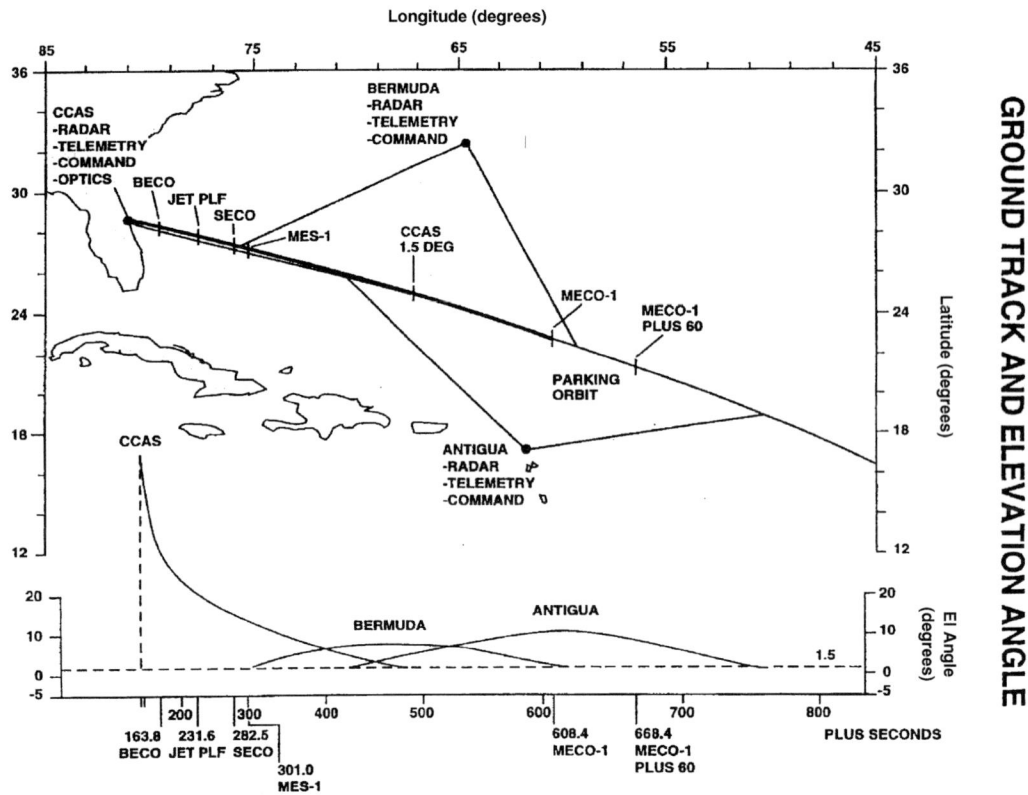

BECO = booster engine cutoff
CCAS = Cape Canaveral Air Station
JET PLF = jettison payload fairing

MECO-1 = (upper stage) main engine cutoff (first cutoff)
MES-1 = (upper stage) main engine start (first start)
SECO = sustainer engine cut off

FIGURE 3-3 Ground track and elevation angle for an Atlas IIA launched from Pad 36A at the Eastern Range on an initial flight azimuth of 104 degrees. Source: 45th SW, 1999.

risk-management perspective, it appears that the Africa gates could be moved safely uprange. The combination of vehicle reliability, short time over land, and high speed make it unlikely that moving the Africa gates to within the coverage of uprange assets (i.e., terminating vehicle tracking, telemetry, and FTS coverage beyond approximately 480 seconds) would violate E_c limits or significantly increase E_c. This conclusion should be validated by more detailed analyses covering current and future launch vehicles of interest. If downrange tracking is needed for reasons other than risk management, those requirements should be documented. However, as already noted, the WR has demonstrated that the collective risk standard can be met without tracking, telemetry, or FTS during the later stages of flight.

Primary Recommendation on Risk Management. AFSPC should define objective, consistent risk standards (e.g., casualty expectation, E_c, of 30×10^{-6} and individual risk, P_c, of 1×10^{-6}) and use them as the basis for range safety decisions. Safety procedures based on risk avoidance should be replaced with procedures consistent with the risk management philosophy specified by EWR 127-1. Destruct lines and flight termination system requirements should be defined and implemented in a way that is directly traceable to accepted risk standards.

Finding 3-4. At the Eastern Range, the downrange location of gates and destruct lines and current requirements for downrange coverage by flight termination, telemetry, and tracking systems are not directly related to accepted risk standards (e.g., E_c of 30×10^{-6} or P_c of 1×10^{-6}) but to a risk-avoidance policy that discourages the overflight of inhabited landmasses whenever possible. The Western Range implements this policy by constraining the azimuth of orbital launches.

RISK MANAGEMENT APPROACHES TO SAFETY

TABLE 3-2 Probability of Failure vs. Phase for the Atlas IIAS

Phase	Start Time (seconds)	End Time (seconds)	Probability of Failure (per second)
Liftoff	0.00	5.00	0.000884
GLSRB burn	5.00	59.00	0.0000855
ALSRB ignition	59.00	64.00	0.000106
ALSRB burn	64.00	103.40	0.000106
GLSRB jettison	103.40	104.40	0.0000943
ALSRB burn continue	104.40	117.30	0.0000943
ALSRB jettison	117.30	118.30	0.0000741
Booster flight	118.30	164.80	0.0000740
Booster engine cutoff	164.80	165.80	0.00398
Booster engine cutoff to booster package jettison	165.80	168.90	0.0000514
Booster package jettison	168.90	169.90	0.00182
Sustainer flight	169.90	190.90	0.0000514
Payload fairing jettison	190.90	191.90	0.0000507
Sustainer flight (continued)	191.90	282.90	0.0000507
Sustainer engine cutoff	282.90	283.90	0.00000121
Atlas/Centaur separation	283.90	284.90	0.00161
Coast	284.90	301.50	0.000000454
Main engine start (upper stage)	301.50	306.50	0.00153
First Centaur burn	306.50	670.00	0.0000257

GLSRB = ground-lit solid rocket booster
ALSRB = air-lit solid rocket booster
Source: 45th SW, 1999.

Finding 3-5. Moving the Africa gates uprange has the potential to reduce the cost of safety-related downrange assets, decrease the complexity of range safety operations, and reduce launch holds and delays. Moving the Africa gates to within the reach of uprange flight termination, telemetry, and tracking systems is not likely to increase E_c significantly or violate established limits. No known international agreements would preclude moving the gates. Thus, in terms of range safety there is no clear justification for retaining downrange assets at Antigua and Ascension. It may also be feasible to move other gates uprange and further reduce the need for downrange facilities.

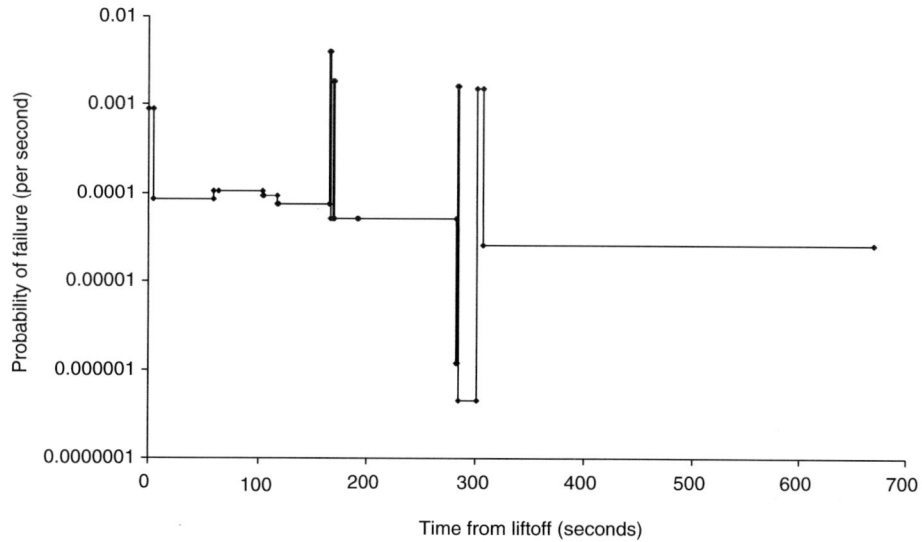

FIGURE 3-4 Probability of failure vs. phase for the Atlas IIAS. Source: 45th SW, 1999.

Primary Recommendation on Africa Gates. While other requirements may exist, from the perspective of launch range safety the Air Force should move the Africa gates to within the limits of uprange flight termination and tracking systems; eliminate the use of assets in Antigua and Ascension for range safety support; and conduct a detailed technical assessment to validate the feasibility of moving other gates uprange. If other requirements for downrange tracking exist, AFSPC should validate those requirements and reexamine this recommendation in light of the additional requirements.

Modeling and Analysis Issues

A number of analytical processes and modeling tools are used before, during, and after launch to predict and monitor the safe operation of launch vehicles. The basic prelaunch safety assessment methodology used at both the WR and ER is based on Monte Carlo simulation tools, which is consistent with methods used in other fields. These tools can compute the likelihood of vehicle failure at a given time during launch, the resulting likelihood of debris impacting a given location, and in some cases the risk of casualties caused by debris impact, explosion, blast, or toxic effects. The final outputs of the assessment are used prior to launch (with statistical wind profiles) to determine whether the launch meets safety criteria and where evacuations are required. Continued evolution and assessment of these modeling and analysis techniques is critical, especially for new types of launch vehicles such as RLVs.

Because of inherent uncertainties in input parameters and modeling assumptions, safety assessments can provide only approximate results, even though range safety personnel are constantly improving their analytical models based on actual range experience. Because the launch rate is quite low compared to the rate at which new technologies are developed, however, it can be difficult to predict the performance of new vehicles or systems using historical data.

Safety procedures and rules should be clearly linked to accepted risk standards, but demonstrating this linkage is difficult because of inherent uncertainties. Therefore, the computation of safety metrics tends to be conservative. For example, models may assume that all debris survives to impact or use worst-case wind profiles. The effect of this conservatism is that actual operations are likely to be safer than predicted. However, conservatism may also overly restrict operations and should be carefully limited. The goal should be to obtain the most accurate answer, not the most conservative one. Safety assessments should be conducted to the level of detail appropriate to the scale and accuracy of the assumptions used in the models, and making the models more detailed is not always warranted. In fact, the additional complexity may have no effect—or even a negative effect—on accuracy.

The results of safety assessments should not be subjectively altered when making decisions regarding launch, evacuation, or flight termination. This issue has been identified and discussed by the RCC:

> Answers obtained by applying these analytical methods . . . are not the "absolute truth" but are the product of a rational process to establish objective safety recommendations. Therefore, the answers should not be subjectively altered at the end of the process. Such changes could render invalid the informed decision process which helps protect the government from liability (RCC, 1997a).

This issue has been further developed in the *Flight Safety Analyst Handbook*:

> While the validity of the calculation process (the "math") is not often questioned, numerical results at the end of the process are sometimes called "ballpark", or thought to contain "margins" as the result of "conservatism built into models." This intuition fosters a belief that numbers indicating high risk (especially borderline high risk) can somehow be discounted. . . . If the Commander exceeds the criteria it can be argued that the criteria does not exist, or was in fact, never a valid criteria. For credibility and liability protection, it is better to change the criteria before initiating a launch operation, than to establish one and then violate it (30th SW, 1999).

As indicated above and in Appendix E, the committee noted that conservatism is, in fact, built into many range safety analytical models and procedures. Ensuring that safety analyses are accurate and free of unnecessary conservatism will help minimize the temptation to discount their results.

Recommendation 3-2. AFSPC should identify and correct unwarranted conservatism in analytical models and verify that modeling and analytical methods are properly implemented. Periodic, independent reviews should be conducted to ensure that the level of modeling detail is appropriate given the accuracy of model inputs and assumptions.

Differences between WR and ER Analysis Methods

The overall modeling and analysis approaches at ER and WR are similar, but some significant differences exist. For example, the ranges use different assumptions and models in computing safety metrics, such as E_c, and they use analytical results differently. At the WR, safety analyses are rerun on launch days using measured wind data to reevaluate the safety metrics and verify that the launch meets the accepted safety criteria. At the ER, the measured winds are compared against predefined worst-case winds to determine if the launch may proceed.

Because of the potential risk to the launch area it is necessary to detect and terminate the flight of a vehicle that fails to pitch over and head downrange. Both ranges compute how long it would take for a vehicle to present an unacceptable risk if it flew straight up. If a vehicle fails to turn downrange by the specified time the flight is terminated. Also, at the

ER, a "chevron display" is used to track the IIP immediately after launch. If the IIP fails to move downrange at the proper rate as shown on the chevron display, destruct commands are sent. At WR, a dedicated pitch-program display is used to track the vehicle's position relative to the nominal programming trajectory. Prior to launch, the WR also computes how long after launch it takes a vehicle to generate enough kinetic energy to impact a region outside the ILLs. If tracking of the vehicle is not be available by that time, the flight is terminated. Differences in the assumptions and methods used at the ER and WR to determine ship and aircraft exclusion zones are discussed further in Chapter 5.

Finding 3-6. The overall modeling and analysis approaches at the Eastern and Western Ranges are similar, but there are some significant differences in analytical tools, assumptions, and operational procedures. These include differences in analysis software packages, methods of defining ship exclusion zones, and displays for monitoring the launch vehicle trajectory. The differences may increase costs because of overlap or duplication of effort in developing models, software, and hardware for the two ranges.

Although differences in geography and other factors may make it impractical for the ER and WR to use identical modeling and analysis approaches, an effort should be made to increase the degree of commonality in accordance with the Primary Recommendation on EWR 127-1.

REFERENCES

30th SW (30th Space Wing). 1999. 30th SW/SE Flight Safety Analyst Handbook. Vandenberg Air Force Base, Calif.: 30th Space Wing.

45th SW (45th Space Wing). 1999. Eastern Range and Western Range Collective Risk and Associated Data. August 13, 1999. Patrick Air Force Base, Fla.: 45th Space Wing.

AFSPC/AFMC (Air Force Space Command/ Air Force Materiel Command). 1997. AFSPC/AFMC Memorandum of Agreement on Spacelift Roles and Responsibilities. January 31, 1997. Peterson Air Force Base, Colo.: Air Force Space Command.

AFSPC/AFMC. 1999. AFSPC/AFMC Memorandum of Agreement on Spacelift Roles and Responsibilities (Draft). Revision 2. May 1, 1999. Peterson Air Force Base, Colo.: Air Force Space Command.

AFMC (Air Force Materiel Command). 1998. Single Manager Roles and Responsibilities. AFMC Pamphlet 63-3, September 1, 1998. Wright Patterson Air Force Base, Ohio: Air Force Materiel Command.

DoD (Department of Defense). 1998. Use, Management, and Operation of DoD Major Ranges and Test Facilities. Department of Defense Directive 3200.11. January 26, 1998.

EWR 127-1 (Eastern and Western Range Safety Requirements). 1997. Available on line at: *http://www.pafb.af.mil/45sw/rangesafety/ewr97.htm* January 20, 2000.

LMA (Lockheed Martin Astronautics). 1999. Atlas II/IIAS Flight Data Package: GTO Class Missions Report No. LMA-AFD-98-290. May 1999. Denver, Colo.: Lockheed Martin Astronautics.

NTSB (National Transportation Safety Board). 2000. Aviation Accident Statistics. Available on line at: *http://www.ntsb.gov/aviation/Stats.htm* January 5, 2000.

RCC. 1997a. Supplement to Common Risk Criteria for National Test Ranges: Inert Debris. Supplement to RCC Standard 321-97. AD-A324955. February 1997. Available on line at: *http://www.jcte.jcs.mil/RCC/manuals/321/index.htm* January 21, 2000.

RCC (Range Commanders Council). 1997b. Common Risk Criteria for National Test Ranges: Inert Debris. AD-A324356. February 1997. Available on line at: *http://www.jcte.jcs.mil/RCC/manuals/321/index.html* January 21, 2000.

USAF (U.S. Air Force). 1991. Air Force Instruction (AFI) 91-202 AFSPC Sup 1. February 1, 1991. Washington, D.C.: U.S. Air Force.

Ward, J. 1997. Estimation of Downrange Risks for Northeast Titan and Athena Launches. Memo for the Record. October 31, 1997. Research Triangle Park, N.C.: Research Triangle Institute.

4

Flight Safety Requirements

One of the most important safety responsibilities of the range commanders (i.e., the commanders of the 30th and 45th Space Wings) is to ensure public safety during launch and flight. Range safety personnel evaluate vehicle design, manufacture, and installation prior to launch; monitor vehicle and environmental conditions during countdown; monitor the track of vehicles during flight; and, if necessary, terminate the flight of malfunctioning vehicles. The method used for flight termination depends on the vehicle, the stage of flight, and other circumstances of the failure. In all cases, propulsion is terminated. In addition, the vehicle may be destroyed to disperse propellants before surface impact, or it may be kept intact to minimize the dispersion of solid debris. Flight termination can also be initiated automatically by a break-wire or lanyard pull on the vehicle if there is a premature stage separation.

Current FTS practices have an excellent safety record. From 1988 through November 1999 there were 427 launches at the ER, during which 11 destruct commands were issued (two Atlas II, one Delta III, one Titan IV, four Trident SLBMs, and three other missiles). Over the same time period there were 177 launches at the WR, during which 11 destruct commands were sent (one Athena, two Pegasus, one Titan IV, and seven ICBMs). Total failure of an FTS is extremely rare at either range, and destruct commands are often superfluous because vehicles explode or break up because of dynamic forces before the mission flight control officer (MFCO) can react.

This chapter discusses current and future requirements for flight termination, tracking, and telemetry; examines cost, reliability, and efficiency from the perspectives of both the ranges and the users; and suggests improvements.

TRACKING

A large fraction of range support costs are related to developing, maintaining, and operating accurate and reliable tracking systems. EWR 127-1 requires "at least two adequate and independent instrumentation data sources" for tracking launch vehicles "from T-0 throughout each phase of powered flight up to the end of range safety responsibility" (Paragraph 2.5.4). For space launch vehicles, the ER implements this requirement by mandating two independent tracking sources and full FTS capability from launch through normal engine shutdown subsequent to achieving orbit. The WR requires two independent tracking sources and FTS capability until loss of contact with the vehicle as it approaches and passes over the horizon. Missions at both ranges are allowed to proceed if one source of data is lost during flight, but complete loss of tracking data is a *prima facie* reason to terminate the flight even if there are no indications that the vehicle is departing from its intended flight path. Thus, the purpose of the two-source requirement is largely to ensure a successful mission—by ruling out the possibility that a good flight will be terminated because a single tracking system fails. Public safety, however, is not based on mission assurance. Safety is provided by the ability to determine when something has gone wrong and, if necessary, safely terminate a flight.

Tracking requirements at both ranges are met through a combination of C-band radar beacons, unaided radar tracking, optics, and vehicle telemetry. Launch vehicle systems include C-band transponders with omnidirectional antennas; two or more flight termination command receivers per vehicle; redundant batteries to power the flight termination receivers; and thrust-termination and vehicle-destruct ordnance, including initiators and safe/arm devices. A real-time vehicle telemetry system also is required to provide telemetered inertial guidance (TMIG) data from the vehicle guidance system, along with other critical data, such as the status of the flight termination receiver and chamber pressures in the engines and solid rocket motors.

Accuracy Requirements

Tracking systems must be accurate, timely, and reliable enough to enable the ranges to calculate accurate IIPs.

EWR 127-1 (Paragraph 2.5.4.1.1) establishes different IIP error standards for the WR and ER. The ER believes that the current limits are overly restrictive and has proposed relaxing them in the next version of EWR 127-1 (Campbell, 1999). For the ER, crossrange and downrange error in IIP must be no more than 100 feet (three sigma)[1] until IIP clears the launch area (as currently authorized) or no more than 300 feet (per the proposed change). The proposed change also requires that error in vehicle position not exceed 140 feet in the vicinity of the launch area. At the WR, the allowable error is 1,000 feet. Once the vehicle clears the launch area, IIP crossrange error at the ER must not exceed 0.5 percent of the IIP range, and downrange error may not exceed 5 percent of the IIP range. At the WR, crossrange and downrange errors must not exceed 1 percent of the IIP range. The ER bases estimates of IIP error on static error sources, excluding time lags caused by data transfer and processing delays. The WR incorporates lag errors in its methodology. This may explain the large differences in accuracy requirements between the ranges.

It is essential to account for lag errors in predicting the accuracy of IIPs because range radars cannot satisfy the IIP accuracy requirements in real time. In addition to mechanical pointing lags and various delays in relaying the data to the ROCC (Range Operations Control Center), sequential-difference data derived from the azimuth, elevation, and range data must be "smoothed" to develop the velocity estimate needed to calculate the IIP. This smoothing typically requires several seconds of data. The accuracy of the real-time IIPs is highly sensitive to the velocity value used, and a sudden change in the vehicle thrust vector takes time for the tracking system to detect and display; meanwhile, the actual IIP may be hundreds of feet from the displayed location.

In practice, however, small errors in IIP are not significant. The debris pattern after an explosive flight termination is many thousands of feet across even for an accident 10 to 15 seconds after liftoff, and pinpointing the center of the pattern is not necessary to ensure safety. The committee supports the proposal to relax the tight IIP accuracy requirements currently imposed at the ER for launch vehicles that have not yet cleared the launch area. Satisfying IIP accuracy requirements outside the launch area is well within the capabilities of both radar and GPS tracking systems.

Radar

The ER and WR each have a network of 10 C-band radars. Several of these radars are located at downrange facilities. Each network consists of one phased-array multiple-object tracking radar and nine, generally aging, C-band single-object tracking radars (see Figure 4-1). Ongoing modernization of both ranges will eliminate the need for most of these radars. As currently planned, the modernized ranges will use differential GPS tracking systems supplemented by two radars at the WR and seven radars at the ER. Three of the seven radars at the ER will be necessary only to support launches of the space shuttle, and three others will be located at downrange facilities to support ballistic missile tests and space object identification.

The WR launches vehicles into polar orbits using initial launch azimuths between 158° and 201°. Downrange assets are not needed for these launches because, by the time uprange facilities lose contact with launch vehicles, they no longer pose a threat to inhabited landmasses.

The ER uses initial launch azimuths of 37° to 114°. Northerly trajectories parallel the U.S. and Canadian coasts. Many launch vehicles on easterly trajectories do not achieve orbit prior to flying over Europe or Africa, and African overflight is common for missions with large payloads headed for geosynchronous orbit. Currently, the ER uses downrange facilities to track vehicles to orbital insertion.[2]

EWR 127-1 specifies that the ground segment of the tracking system must have a reliability of at least 0.999 for a one-hour duration during the period of range safety responsibility (Paragraph 2.5.4.1.3). EWR 127-1 also says that the reliability requirements for vehicle-based range tracking systems are 0.995 for the C-band transponder systems and 0.999 for GPS-based systems (Paragraph 4.10.3a). A proposed change to EWR 127-1 would establish a slightly lower reliability standard (of 0.96) for each of the two independent sources of tracking data used in a GPS-based system (Cather, 1999).[3] The requirement to have two independent tracking sources would not be changed.

TELEMETRY

Telemetered data are routinely collected during launches, and selected items are provided in real time to the range user and to range safety personnel in the ROCC. Data of particular interest to range safety are guidance data, command receiver status, and steering commands.

TMIG data must be used as a tracking source for launch vehicles equipped with an inertial guidance system (EWR 127-1, Section 2.5.5.1). TMIG data are important because they provide vehicle state vectors (which indicate vehicle location and velocity) to determine IIPs with a minimum of data processing. Compared to IIP displays based on

---

[1]*Sigma* is a measure of statistical fluctuation. *Three sigma* means that the probability that the outcome will fall within expected limits (in this case, the probability of having an IIP error of 100 feet or less) should be at least 99.7 percent. The equivalent values for one and two sigma are 68 and 95 percent.

[2]As discussed in Chapter 3, the committee recommends that the Air Force modify this practice because GPS metric tracking will eliminate range safety requirements for downrange radars for space launches.

[3]As noted below, the two sources of tracking data used with a GPS system would be (1) a GPS device (translator or receiver) and (2) another, independent GPS device, an inertial measurement unit (IMU), or TMIG.

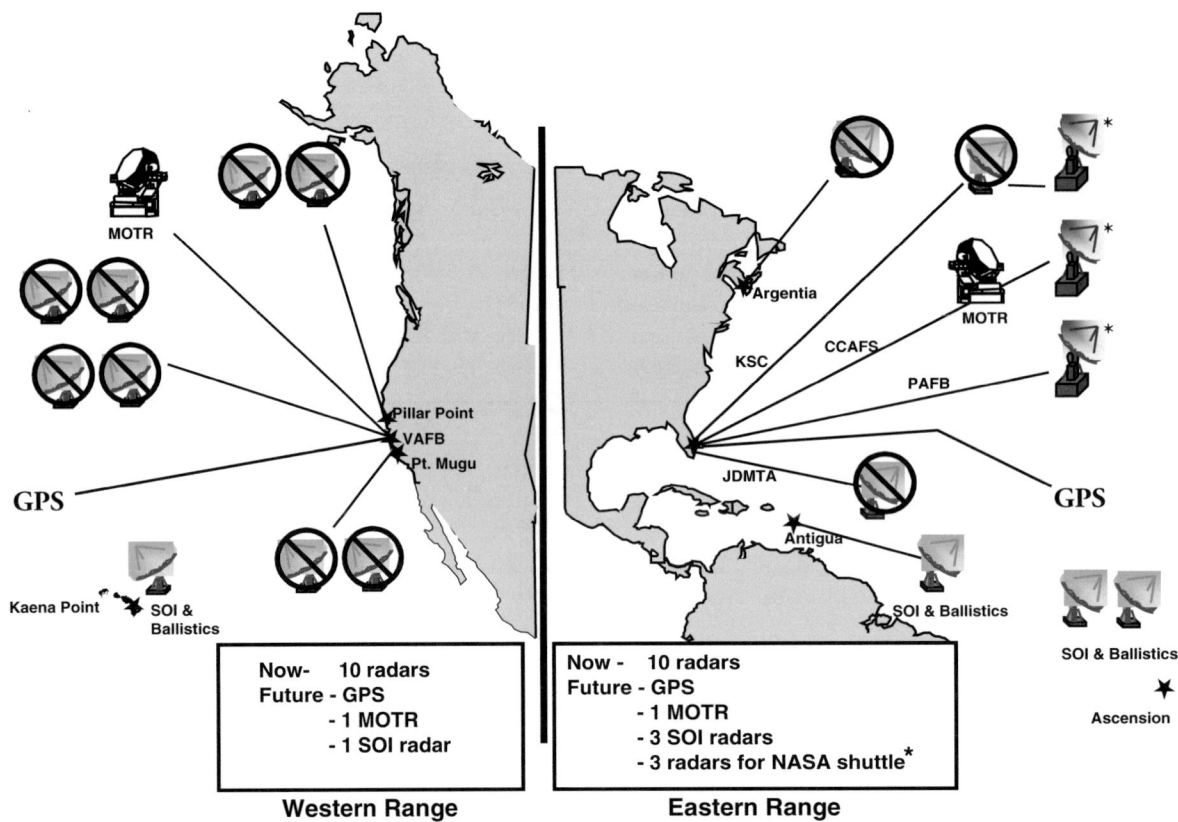

CCAFS = Cape Canaveral Air Force Station
JDMTA = Jonathan Dickinson Missile Tracking Annex
KSC = Kennedy Space Center
MOTR = multiple-object tracking radar

PAFB = Patrick Air Force Base
SOI = space object identification (used to determine satellite size, shape, motion, orientation and operational status)
VAFB = Vandenberg Air Force Base

FIGURE 4-1 Changes in range tracking support under the RSA range modernization program. Source: Finn and Woods, 1999.

data from a single radar, IIP displays based on TMIG data are very low in noise, extremely accurate, and have very little delay.

TMIG data are also useful during tracking system dropouts or overshoots, which occur frequently at major staging events. TMIG data can quickly indicate sudden turns or rotations, and a failure during a staging event can be detected using TMIG data three or four seconds earlier than with radar data. This may be less important with GPS tracking systems, which are expected to reduce lag times. Even so, GPS data links may be interrupted for a second or two by staging events; flight experience will be necessary to determine the effect of staging events more accurately.

Inertial guidance systems can fail in a way that would steer a vehicle off course even though TMIG data indicate that the flight is on course. For this reason, TMIG data must be independently verified by other sources early in flight. At the WR, MFCOs have used TMIG to track ICBMs for three decades, but only after visual comparison of TMIG data with radar data through the beginning of fourth-stage flight. At the ER, MFCOs shift to TMIG as a prime source on SLBM missions when other tracking data becomes unreliable.

At the WR there have been 16 failures involving TMIG during tests of Minuteman III ICBMs. In two cases, TMIG data apparently failed to indicate an off-course condition (Cortopassi, 1999). The first was W7262 launched on January 26, 1972, which missed the Kwajalein target point by five miles because of a human error in prelaunch alignment. The resulting offset early in the trajectory was too small to be detected in real time by radar. In a second incident in March 1990, the TMIG did not show on-course progress, but it indicated a safe condition when, in fact, the IIP was rapidly moving off the nominal flight path. More recently, on August 12, 1998, a Titan IV failure at the ER involved a momentary power failure and reset after the vehicle had pitched over to head downrange. The reset apparently caused the guidance system to act as if the vehicle were still level on the pad; consequently, it reinitiated pitch steering. The vehicle was destroyed by aerodynamic forces (Lyles, 1999).

None of the events just described led to an uncontrolled situation. That would have required multiple independent failures, such as:

- the loss of primary tracking so that TMIG data become the primary source of tracking data
- a guidance system failure causing the vehicle to turn to a prohibited azimuth
- a turn ending before aerodynamic limits have been exceeded, with return to stable flight
- the absence of any other observations or information indicating the vehicle is off course

The probability that all of these conditions would occur on a single launch is very small. Therefore, the committee believes that TMIG should be allowed to serve as one of the two required sources of tracking data, and the process of verifying TMIG data against other tracking sources should be retained.

Vehicle operators normally have their own requirements for real-time telemetry displays, and only minor changes (and minimal costs) are required to strip out and forward items of interest to the range safety office. Current and planned telemetry equipment can readily handle the various formats, reporting rates, and other differences between launch vehicles and real-time displays.

Stand-alone inertial measurement units (IMUs) also can be used as independent tracking sources. IMUs must be calibrated prior to launch. Also, unless an IMU is connected to a GPS unit or can be otherwise recalibrated during flight, the IMU must include gyros with very low drift rates. Gyros with low drift rates are very expensive, and an acceptable stand-alone IMU might cost almost as much as a full guidance system. However, it is quite feasible to pair a GPS receiver with an IMU costing on the order of $10,000.

Recommendation 4-1. As a matter of good engineering practice, the requirement for two independent sources of tracking data should be retained, and the accuracy of telemetered inertial guidance data should be verified after launch. AFSPC should clarify EWR 127-1 to specify that telemetered inertial guidance data can serve as one of the two sources of tracking data.

GPS METRIC TRACKING

As part of the RSA range modernization program, the Air Force plans to transition both the ER and WR to GPS-based tracking systems. This will affect both range operators and users.

GPS System Options

A GPS system would replace the onboard C-band transponder beacon with either a GPS translator or receiver unit along with appropriate cabling and L-band antennas. Ground-based radars would be replaced with telemetry-receiving equipment compatible with the chosen system.

A GPS translator receives L-band signals from GPS satellites and retransmits them without any processing to the ground on the S-band telemetry link. The alternative, a GPS receiver, would use L-bank signals to calculate vehicle position and state vectors and transmit them to the ground in real time. In both cases, data from the satellite are transmitted to the ground via an S-band communications link and then relayed to the ROCC. The accuracy of translators or receivers can be improved through the use of differential GPS corrections, which employ a fixed receiver on the ground as a reference to account for errors associated with (1) intentional inaccuracies in the signals transmitted by the GPS satellites, (2) small deviations in the orbits of the GPS satellites, and (3) atmospheric effects that distort the GPS signals received by the launch vehicle. Both differential GPS receiver and translator systems, if properly designed and qualified, would be able to meet range requirements for tracking accuracy.

Analog GPS Translator System

Launches of Navy SLBMs conducted offshore from Cape Canaveral use a GPS-based analog translator system for range safety. Simpler versions of this system were used for SLBM launches beginning in 1979. Since the completion of the GPS constellation, the system has not constrained launch timing or flight trajectory. The L-band GPS signals are captured by the vehicle, translated into an S-band transmission, and relayed through appropriate ground-based telemetry receivers to the ROCC. The ground-based receivers are configured to decode the S-band transmission and add correcting algorithms, producing highly accurate vehicle state vectors. The overall accuracy of this system exceeds current requirements, and the system is fully flight operational. However, the GPS analog translator system has some drawbacks. First, a wide bandwidth, in the neighborhood of several megahertz, is required to downlink the GPS signals to the range receivers. This creates a problem because the S-band telemetry spectrum is already overloaded, as are the communications links between downrange sites and the ROCC. Second, noise increases with bandwidth, so more power is needed on the vehicle for an acceptable signal-to-noise ratio. Also, noise is added each time the signal is retransmitted, which may prevent ground-based receivers from locking on to the signals or decoding them in real time.

The ER already operates GPS ground-processing stations to support the launch of Navy SLBMs. As described below, the WR is procuring and installing GPS ground-processing stations capable of supporting both analog and digital GPS translator systems to support ICBM launches.

Digital GPS Translator System

The Air Force Range Instrumentation System Program Office at Eglin Air Force Base is developing a digital GPS translator system, which translates and transmits L-band GPS signals to the ground in an S-band digital format. Ground processing of the signals to determine vehicle position and velocity are the same as in the analog system after the digital telemetry is decoded. In fact, the current version of the ground translator processor station can process either analog or digitized signals equally well.

Compared to an analog system, the digital system reduces the S-band retransmission bandwidth by a factor of 2 to 10 and reduces the size and weight of the onboard components. The digitized signal provides more robust link margins and facilitates signal relaying, while the reduced transmission bandwidth reduces interference with existing and planned launch vehicle telemetry. These units are being adapted specifically for launches of Air Force ICBMs and are not presently planned or being demonstrated for space launches. Flight qualification is scheduled for completion in early 2000.

GPS Receiver System

GPS receiver systems have been flown experimentally on vehicles at the U.S. Army White Sands Missile Range, at Vandenberg Air Force Base, and on Pegasus vehicles launched on both coasts. The development of an operational GPS receiver for use at the ER and WR is under way. Designs under consideration would enable simultaneous tracking of up to 12 GPS satellites. The vehicle state vector and the intermediate information used to determine the state vector would be transmitted to the ground. The RSA ground support system will include fixed GPS receivers to produce a differentially corrected state vector. The S-band bandwidth necessary to send receiver data to the ground is on the order of 100 times smaller than the bandwidth for a translator system.

Precise estimates of the accuracy and precision of GPS-derived trajectory information must take into account antenna performance, how well corrections can be made for refraction and other errors, the number of satellites used to generate the solution, and other factors. Without differential correction, real-time position accuracy of 500 feet (one sigma) is routinely obtained by receivers used in boats and automobiles. Commercially available differential GPS receivers can provide positional accuracy of significantly less than 100 feet, which is the most stringent accuracy requirement at either of the ranges. GPS systems have demonstrated the ability to maintain lock on the satellite signals at accelerations significantly higher than those expected during booster flight of space launch vehicles. Also, GPS receiver systems can be linked with inexpensive auxiliary IMUs to compensate for momentary loss of signals resulting from staging or other dynamic events. Necessary algorithms have already been developed, and experimental testing has demonstrated their versatility and robustness over a broad range of simulated flight conditions.

Assessing the Alternatives

Costs

Switching to a GPS-based range safety tracking system would mean replacing each onboard C-band transponder with either a GPS receiver system or a GPS translator system with an S-band telemetry transmitter separate from the one used for TMIG data. The digital translator developed for ICBMs is projected to cost roughly $28,000 each (Wells, 1999). The launch-hardened G-12 GPS receiver built by Ashtech being tested at White Sands and Edwards Air Force Base (but is not space qualified) reportedly costs between $25,000 and $30,000. A receiver being built by Rockwell-Collins Radio for the ICBM terminal area study of the Reentry Vehicle Decoy GPS Experiment is projected to cost $10,000 to $12,000 and will be tested early in 2000 (Finn and Woods, 1999). These devices can track the required number of GPS satellites, and their output data rate satisfies requirements for space launch. However, the performance of the receivers has not yet been validated in terms of data time lags, compatibility with the vibration and temperature environment of flight, and other factors listed in Chapter 4 of EWR 127-1.

The cost and time required to develop a GPS receiver—and for modifying the design of launch vehicle hardware and interfaces to make them compatible with GPS receivers—are still uncertain. SMC estimates it will cost $5 to $10 million to redesign and recertify each family of launch vehicles to use GPS receivers in place of C-band transponders (Finn and Woods, 1999). Some users are concerned that actual costs could be significantly higher. To meet schedule requirements during the transition, existing radars should remain functional long enough to acquire GPS receivers, modify the design of launch vehicles, and conduct flight tests (including operational tests during which users, if they wish, can fly GPS receivers in parallel with traditional radar tracking systems to build confidence in the new systems).

Shifting from radar tracking to GPS receivers will also increase recurring costs for related vehicle systems. The added L-band antennas, L-band low-noise amplifiers, cables, and the GPS receiver/transmitter could double the cost of onboard hardware compared to a C-band transponder system (Smith, 1999). Depending on performance and configuration requirements, the first operational designs could add 10 pounds or more to the weight of the upper stage, resulting in an equivalent reduction in payload capacity. Even so, the life-cycle costs for a GPS receiver tracking system are expected to be much lower than for radar systems because the current single-object radars are aging, expensive to maintain, and would be expensive to replace (see below).

Versatility

Both GPS receivers and translators can be designed and built to meet basic safety requirements related to qualification tests, data rate, data latency, overall accuracy, and reliability. The performance of both systems is sensitive to the design and placement of the L-band antennas and would differ somewhat (but probably not materially) in the recovery time from momentary dropouts in the L-band signals received from GPS satellites or S-band transmissions to the ground. As already noted, GPS translators require a broader bandwidth and may experience earlier loss of signal lock than receiver systems. Also, if a vehicle equipped with a translator system departs from the planned flight path, flight termination would have to be commanded from the ground; the GPS translator system would not detect the off-course condition because GPS position data would not be available on board.

GPS receivers, which would compute position and velocity data on board, would enable development of FTSs with a higher degree of autonomy. Also, the small bandwidth requirements of receiver systems would be compatible with a space-based range that would use satellites to relay commands, GPS data, and telemetry between the vehicle and the range control center. A space-based range would eliminate the need for virtually all ground-based tracking infrastructure, greatly reduce the need for ground-based telemetry equipment, and allow a control center to support launches almost anywhere in the world. Thus, a GPS receiver system offers operational advantages over either a translator system or the existing radar-based system.

RANGE MODERNIZATION

User costs at both the ER and WR are based, in part, on the ranges' direct costs of supporting a particular launch (i.e., how much each launch increases the overall cost of range operations). Lockheed Martin Astronautics estimates that the average range costs for each Atlas or Delta launch are on the order of $500,000. In addition, prelaunch tests of the FTS cost $100,000 and tests of the C-band radar tracking beacon and S-band vehicle telemetry system cost another $100,000 (Hillyer, 1999). These costs do not include the cost of standard factory acceptance tests. Lockheed Martin estimates that a "typical" mission scrub costs the range user about $110,000 (Hillyer, 1999). This cost is highly variable, however, depending on the cause of the scrub, how long it takes to reschedule the launch, and the effect of the scrub on the flight status of the vehicle and payload systems.

Of greater concern is the need to upgrade or replace range radars. The vast majority of existing radars were built in the early 1950s, and their continued use will increase the already high cost of maintaining their aging electronics and pointing mechanisms. Increasing age is also expected to increase the failure rate of critical systems and down time for repairs. Increases in average and peak launch rates and shorter launch windows will amplify the impact of unexpected ground system down time and make it more difficult to meet users' schedule requirements. During the 196 launches at the ER from 1993 through 1998, problems with range instrumentation caused numerous holds but only five launch scrubs. Over the same time period, the WR conducted 92 launches, and instrumentation problems caused three scrubs.

As already discussed, a GPS-based range safety tracking system would eliminate the need for 11 of the 20 tracking radars currently used to support launches at the ER and WR. Studies initiated by SMC show that a GPS-based system would significantly reduce total life-cycle costs (LMTSC, 1996, and Finn and Woods, 1999). Figure 4-2 shows estimates of life-cycle cost for three options: (1) retaining current radars with no major upgrades, (2) modernizing current radars and adding two new ones, and (3) transitioning to a GPS tracking system supplemented by nine radars. The estimates include development costs (of $30 to $60 million) and recurring costs (of $120,000 to $250,000 per vehicle) of implementing a GPS system on all current launch vehicles except the space shuttle. The results do not take into account potential future savings from the elimination of scrubs caused by radar outages, reduced radar infrastructure, easier prelaunch checkout, or the cost to NASA of either modifying the space shuttle fleet to use GPS (estimated by NASA at $32 million) or maintaining a radar tracking capability solely for shuttle operations.

Most of the savings that are explicitly included in Figure 4-2 are in the area of ground system acquisition and maintenance. Because of the way costs are allocated, these savings would primarily benefit the range owner (the Air Force), not users (industry). Costs might also be reduced through collaborative development of new flight safety systems. Until the 1970s the government designed and certified many safety systems and provided them to launch contractors as government-supplied equipment. The committee believes that the time and total cost of shifting to a GPS receiver tracking system could be reduced if government and industry work together to develop and certify new systems that satisfy agreed-upon performance requirements and are compatible with a broad range of launch vehicles in terms of weight, size, and power requirements.

The schedule for deploying GPS receiver tracking systems (i.e., completing the RSA program and related activities) is based primarily on the amount of available funding. In recent years annual funding has been reduced, which has delayed the schedule and increased total program costs. The committee believes it would be worthwhile to restore funding and accelerate the deployment of GPS receiver tracking systems because of the cost savings and operational improvements that would result.

Finding 4-1. For space launches, an onboard GPS receiver tracking system would be more versatile and have lower total life-cycle costs than GPS translator or radar tracking systems.

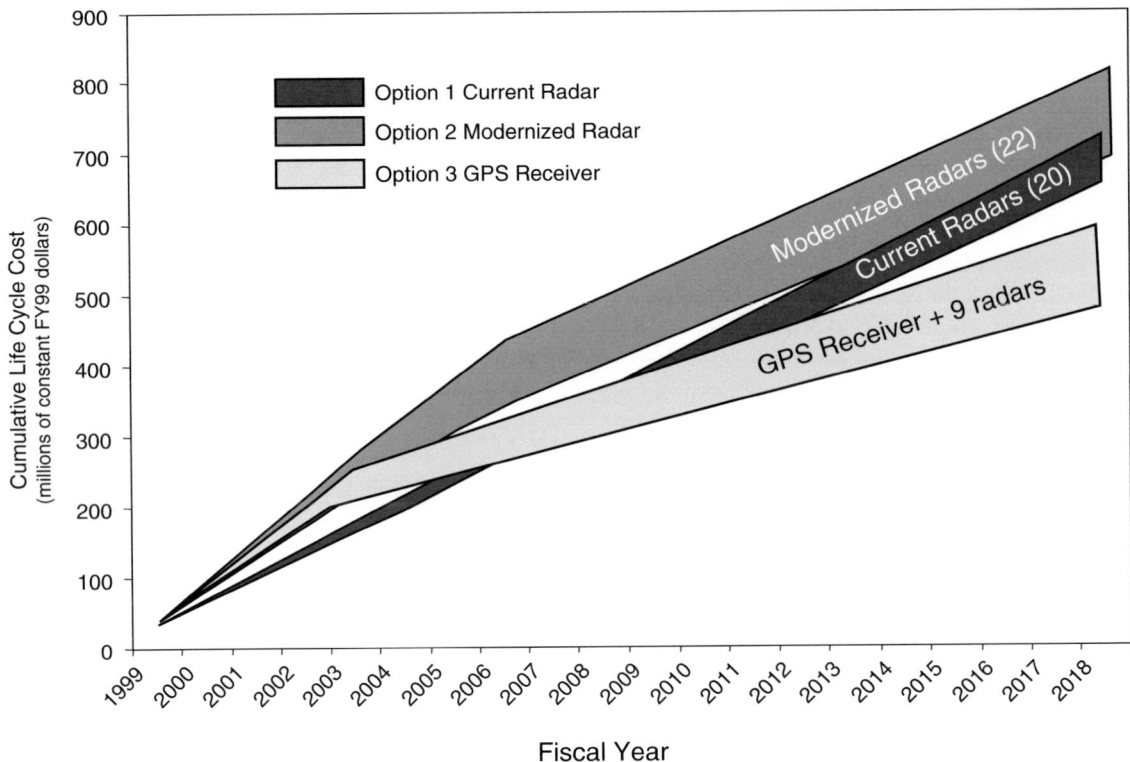

FIGURE 4-2 Comparison of life-cycle costs for radar and GPS-based range tracking systems. Source: Finn and Woods, 1999.

Finding 4-2. Real-time GPS tracking systems have an overall cost and performance advantage over the single-object radar network that has been the workhorse on both the Eastern and Western Ranges for many years. Implementation of a GPS tracking system would increase users' recurring and nonrecurring costs in the short term, but it would benefit users in the long term by increasing operational flexibility. A GPS tracking system would also yield long-term costs savings for the ranges.

Primary Recommendation on GPS Receivers. AFSPC should deploy a GPS receiver tracking system as the baseline range tracking system for space launch vehicles. The transition to GPS-based tracking should be completed as rapidly as feasible.

Finding 4-3. Upgrades to onboard tracking systems currently in use and to new systems, such as GPS receivers, are relatively costly for individual users. Each user currently must develop or acquire hardware, prove that it meets safety requirements, demonstrate its compatibility with range support equipment, provide for qualification and acceptance testing, and support confidence checks in the final countdown.

Recommendation 4-2. AFSPC should form a range-industry team to define performance requirements and technical specifications for the onboard elements of a GPS receiver tracking system, including cost, weight, size, and power limitations, and to establish user requirements during the transition from radar to GPS-based tracking systems. A cost-shared government/industry project should be established for the development and qualification testing of common end-user equipment. Range users should pay for the recurring costs of onboard hardware.

After the committee had completed its deliberations, representatives of Lockheed Martin Astronautics and the Boeing Company gave the committee copies of an assessment that they had performed on alternative approaches to range modernization. Their assessment concluded that the most cost-effective option would be to use a dual TMIG system in place of a radar or GPS tracking system. Although the committee did not have an opportunity to examine this assessment in detail, some of its assumptions and cost projections seemed questionable. The information received by the committee was not convincing and the committee stands by its recommendation that the ER and WR should be modernized with GPS receiver tracking systems for space launch.

AUTONOMOUS FLIGHT TERMINATION SYSTEMS WITH GPS

GPS-receiver range tracking systems would be compatible with either a fully autonomous FTS or a conventional, human-in-the-loop FTS with enhanced autonomous functions. Some autonomous functions are already operational in the form of inadvertent-separation destruct systems that sense unplanned vehicle breakup and initiate the destruct sequence.

At least three basic methods could be used to implement an autonomous FTS that would assume full FTS responsibility once the launch vehicle clears the launch area. One method is to continuously compute the flight heading, compare it with the expected heading, and terminate the flight if the difference exceeds a predetermined limit. This method could use headings derived from the GPS receiver system, the vehicle guidance system, or an auxiliary IMU.

A second method is to continuously compute position and velocity and compare them with expected values. The flight would be terminated if errors exceeded predetermined limits as a function of flight time. This would be a relatively simple and straightforward way to use GPS data.

A third method is to replicate the current IIP and surface destruct line methodology. This is the most complex of the three methods described here, but it would provide the greatest margin for allowing continued flight of a vehicle that deviates from the intended flight plan but is not yet dangerous. Allowing a flight to proceed under such circumstances provides an opportunity for the vehicle to recover and maximizes the collection of telemetry to support postmission investigations of the problem.

Each of these methods would require a logic unit with validated software to process measured data, such as GPS state vectors; compare them with stored data; and generate commands to the FTS. Like current FTS flight hardware, autonomous FTSs would be designed for high reliability, tested for compatibility with the launch environment, and include means for prelaunch verification of operability.

Certain Russian launch vehicles use fully autonomous FTSs, although they do not use GPS (or equivalent) systems. In one example, projected allowable pitch and yaw angle limits based on time from liftoff are stored in the onboard computer and compared to actual launch vehicle flight angles from the guidance IMU. Figure 4-3 shows the limits for a typical Sea Launch mission using a Ukrainian/Russian booster system. The Sea Launch system does not satisfy ER or WR requirements for two independent data sources, but it does demonstrate the feasibility of using autonomous FTSs.

The Israeli Arrow missile uses an automated FTS during portions of flight where the allowable flight path is quite narrow and human reaction time may not be fast enough to terminate the flight of a malfunctioning rocket safely. One Arrow was destroyed by its FTS during a test flight, although some believe the termination could have been avoided if a human operator had been in the primary control loop.

At locations such as the ER, the uprange launch area is the largest contributor to collective risk. An initial approach to autonomous FTS would be to fly the uprange portion of the mission using traditional human-in-the-loop FTS procedures. Then, after E_c has decreased with distance and flight time, the mission could switch to a fully autonomous FTS before leaving the uprange area. A semiautonomous FTS such as this would probably require minimal hardware modifications to the GPS receiver. Retransmission of the GPS-derived position and velocity data on independent S-band links would no longer be necessary, the need for downrange facilities would be further reduced, and practically all uprange tracking and telemetry assets could be eliminated, resulting in very substantial cost savings. The committee believes that the reliability of such a system could be demonstrated to equal or exceed current FTS system requirements.

Although an autonomous FTS system is technically feasible, system performance requirements must be defined, development and validation costs must be accurately estimated, and issues of public acceptability (domestically and internationally) must be addressed to determine if a fully autonomous FTS would be practical and cost effective. The successful deployment of semiautonomous systems, which would provide operational benefits even if a fully autonomous system is never developed, would help resolve these issues. A first step would be to develop computerized simulations of vehicle dynamics and FTS responses to determine system requirements. These simulations could be followed by validation testing using sounding rockets or other low-cost test vehicles.

Finding 4-4. With the incorporation of onboard GPS receivers, semiautonomous and fully autonomous flight termination systems would become technically feasible. These systems might substantially reduce range support costs, but additional research and testing is needed to resolve outstanding issues and quantify the likely benefits.

REUSABLE LAUNCH VEHICLES

Many different organizations are developing commercial RLVs, some of which may someday operate from the ER or WR. The basic safety criteria for RLVs should be the same as for expendable launch vehicles in terms of E_c, P_c, and P_i. Even so, range safety processes for RLVs require special attention because RLV concepts vary greatly in design and operational characteristics. In addition, RLVs that carry human beings raise additional safety of flight issues, especially with respect to the acceptability of autonomous or semiautonomous FTSs. FAA regulations and EWR 127-1 should facilitate efforts by range users to obtain launch authorizations using means of compliance that make sense for the

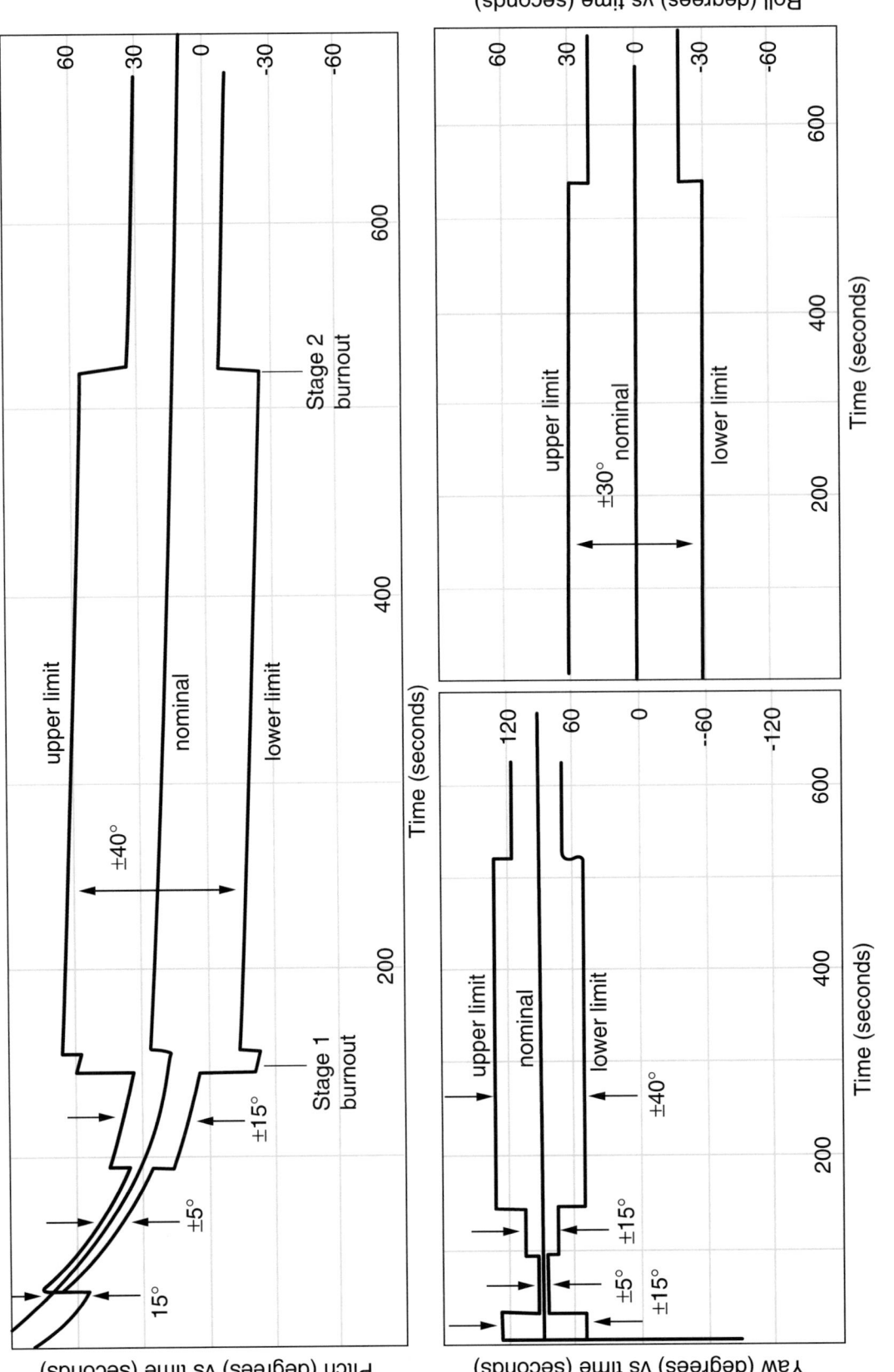

FIGURE 4-3 Flight safety angle limits. Source: ICF Kaiser, 1999, Appendix B.

vehicle being launched. The committee did not attempt to develop specific means of compliance for new classes of RLVs because they will vary with the design of each vehicle. However, streamlining EWR 127-1 to focus on performance-based requirements, in accordance with the Primary Recommendation on EWR 127-1, would be an important first step toward the routine launch of commercial RLVs from either the WR or ER. In the long term, developing space-based ranges would increase operational flexibility and reduce turnaround times. These improvements will be important to support future RLVs, which will also have a high degree of operational flexibility and short turnaround times.

REFERENCES

Campbell, M. 1999. EWR 127-1 Change Request 97-2-009. Submitted by M. Campbell, 45th Space Wing, June 30, 1999. Available on line at: *http://www.pafb.af.mil/45sw/rangesafety/99cr.htm* January 21, 2000.

Cather, C. 1999. EWR 127-1 Change Request 97-4-003. Submitted by C. Cather, 30th Space Wing, April 23, 1999. Available on line at: *http://www.pafb.af.mil/45sw/rangesafety/99cr.htm* January 21, 2000.

Cortopassi, R. 1999. Memorandum from R. Cortopassi, 30 SW/SE, to Director of Safety, 30 SW. Updated Minuteman III History. July 9, 1999.

EWR 127-1 (Eastern and Western Range Safety Requirements). 1997. Available on line at: *http://www.pafb.af.mil/45sw/rangesafety/ewr97.htm* January 20, 2000.

Finn, G., and T. Woods, 1999. Briefing materials on Spacelift Range Metric Tracking Life-Cycle Cost Review. August 10, 1999. Briefing materials prepared by G. Finn, Aerospace Corporation and T. Woods, Air Force Space and Missile Systems Center, in response to a request by the Committee on Space Launch Range Safety, August 13, 1999.

Hillyer, T. 1999. Atlas Space Launch Vehicles—System Cost. Briefing by Tom Hillyer, Lockheed Martin Astronautics, to a panel of the Committee on Space Launch Range Safety, Lockheed Martin Astronautics, Denver, Colorado, June 24, 1999.

ICF Kaiser Consulting Group. 1999. Environmental Assessment for The Sea Launch Project. Prepared for the Federal Aviation Administration, February 12, 1999. Available on line at: *http://ast.faa.gov/pdf/sea_launch* January 24, 2000.

LMTSC (Lockheed Martin Tactical Systems Company). Global Positioning System Trade Study. August 1996. Fort Worth, Texas: Lockheed Martin Tactical Systems Company.

Lyles, L. 1999. U.S. Department of Defense Assessment of Space Launch Failures. November 1999. Available on line at: *http://www.af.mil/lib/misc/spacebar99.htm* January 21, 2000.

Smith, D. 1999. LMA Comments to GPS Metric Tracking Initiative. Briefing by Dan Smith, Lockheed Martin Astronautics, to a panel of the Committee on Space Launch Range Safety, Lockheed Martin Astronautics, Denver, Colorado, June 24, 1999.

Wells, L. 1999. Translated GPS Range Systems. Briefing by Lawrence Wells, L3 Communications–Interstate Electronics, to the Committee on Space Launch Range Safety, Santa Maria, California, July 12, 1999.

5

Incursions

Aircraft and marine incursions into restricted airspace and waters have not resulted in any casualties and have contributed to only a small percentage of launch holds and scrubs at either the ER or WR. When they do occur, however, delays caused by incursions can be highly disruptive and costly for both the range and the user. Incursions are a particular concern for high profile missions, such as the flight of John Glenn on the space shuttle, that attract large numbers of sightseers, many of whom choose to view the launch from a boat or aircraft. During launch delays caused by intruders other problems may develop, which makes it difficult to assess the historical or future impact of intruders accurately.[1] A launch scrub typically increases launch costs by at least $100,000 and increases risk, especially if it is necessary to de-fuel the launch vehicle or offload the payload.

Space launches typically require a huge investment of time and effort on the part of designers, manufacturers, testers, and support personnel to reduce the likelihood of failures and anomalies. Launch holds triggered by unwitting or irresponsible intruders can frustrate the huge effort that has been made to ensure a successful, on-time launch. Long delays can also adversely affect team performance during subsequent preparations for launch.

Expected increases in marine and airborne traffic, particularly at the ER, and an increase in the pace of space launches, especially commercial launches, are likely to intensify the overall impact of incursions. Increased traffic will further challenge the adequacy of existing surveillance and interdiction capabilities, and a higher launch rate means there will be less time between launches. Thus, not only can the number of scrubs and holds be expected to increase substantially, but their occurrence will be more likely to create a "cascade" effect on subsequent launches, increasing launch disruptions, risks, and costs.

This chapter summarizes the committee's examination of the Air Force's safety guidelines and procedures associated with errant aircraft and surface vehicle incursions into restricted airspace, waters, and terrain. The purpose of this examination was to determine if the number of holds and scrubs could be reduced while desired levels of safety are maintained. The following sections summarize current guidelines and procedures associated with aircraft and marine incursions, describe planned improvements, and recommend further improvements. The ER and the WR use similar equipment and procedures for surveillance and interdiction, and the findings and recommendations that appear in this section generally apply to both ranges.

CURRENT GUIDELINES AND PROCEDURES

As discussed in Chapter 3, launch risk is expressed in terms of E_c (for collective risk), P_c (for individual risk), and P_i (for the probability of hitting a particular ship or aircraft). E_c provides a measure of the overall risk of the launch to the population as a whole; P_c and P_i are used to evaluate the risk to a person or vehicle at a particular location.

The establishment of hazardous launch areas (e.g., flight hazard and flight caution areas) is fundamental to dealing with errant intruders in space launch operations. Hazardous launch areas begin at the launch points and extend downrange along the intended flight azimuths. The size and shape of the areas are based on calculations of P_i for ships and aircraft for the specific launch vehicle, its payload, the known and expected failure modes and effects (including toxic hazards), and weather conditions.

During a launch, only mission-essential personnel are allowed inside flight hazard areas, where the probability that an individual will be seriously injured or killed, P_c, exceeds 1×10^{-5}, and those personnel must take shelter in blast-hardened structures with adequate breathing protection.

---

[1] If an intruder causes a hold, during which an equipment malfunction causes a scrub, the event will probably be recorded as an equipment-induced scrub, and the role of the intruder may not be recorded. Figures obtained from different sources on the number and causes of holds and delays were inconsistent, and the committee was unable to resolve the differences.

Outside each flight hazard area the ranges define a flight caution area, where P_c exceeds 1×10^{-6}. During a launch, only mission support personnel equipped with breathing protection are permitted inside the caution area.

The presence of offshore oil platforms at the WR is a complicating factor with regard to the evacuation of high-risk areas. Oil companies maintain evacuation and other contingency plans for their oil rigs. However, because of concerns about oil spills on unattended oil rigs, even when oil platforms are "evacuated" because of launch risk, the evacuation orders allow a minimum cadre of personnel to remain on board. These difficulties notwithstanding, exclusion may be the only viable option for some high-risk areas.

Risk assessments are used to evaluate risk and define hazardous launch areas. The ER and WR rely on quantitative methods and data to model hazards and to assess risks to people and property. The modeling algorithms and risk calculations used by the ranges seem to conform to generally accepted practices, some of which have been in use for many years. During the initial development and subsequent refinement of these methods, they have been reviewed repeatedly in collaboration with other organizations, including the Lawrence Livermore National Laboratory, NASA, the U.S. Navy, and the National Oceanic and Atmospheric Administration. Therefore, rather than trying to validate the individual analysis methods, the committee assessed more generally the acceptability of the risk standards currently used at the ER and WR to protect the public.

Risk Standards for Aircraft and Ships

As indicated above, the individual risk standard, P_c, for members of the general public is 1×10^{-6}. A different risk standard is appropriate for individual ship-hit probability, P_i, because hitting a ship with a piece of debris will not necessarily result in casualties. At both the WR and ER the maximum allowable P_i is 1×10^{-5}. Chapter 7 of EWR 127-1 (Section 7A.2.8.3.1), however, directs the WR and ER to apply P_i differently. At the WR a launch hold or scrub may be initiated if an individual vessel is exposed to a P_i greater than 10^{-5}. At the ER, however, a launch hold or scrub may be initiated if the *sum* of the individual hit probabilities for all targets plotted within, or predicted to be within, the established probability contours exceeds 1×10^{-5}. Thus, the procedure specified for the ER is more conservative than the procedure for the WR.

Unlike the WR, the ER estimates impact blast overpressure to determine if a near miss of solid propellant fragments debris might damage a ship. That is, if calculations show that debris could miss a vessel but create an overpressure of 0.5 psi (for boats) or 2 psi (for ships), the risk analysis tool counts that as a hit. These limits are not specified in EWR 127-1, and the committee was unable to locate analytical studies supporting the overpressure limits used by the ER, which appear to be quite conservative. However, the committee endorses the use of overpressure limits to improve the accuracy of P_i estimates and recommends adding validated overpressure limits to EWR 127-1.

The number of casualties will depend on the location of the debris impact relative to the people on board and the characteristics of the ship and the debris. In the 30th Space Wing's *Flight Safety Analyst Handbook*, a highly simplified approach (a person sitting on an unprotected deck) indicates that a ship-hit probability of 1×10^{-5} corresponds to an individual risk of 4×10^{-8}. For a launch rate of 33 per year, a P_i of 1×10^{-5} implies that a ship in a particular location would be hit once about every 3,000 years. That figure appears to be in line with an E_c of 30×10^{-6}, which corresponds to three casualties every 3,000 years for the same launch rate. During the study, the committee asked the Air Force to provide a more sophisticated analysis of risks to ships. In response, the Air Force used the Launch Risk Analysis computer program to assess the risk faced by a medium-sized ship with a crew of 10, assuming that the ship resembled a building with light protection characteristics. The results of this analysis, which were provided to the committee, showed that a plot of E_c equal to 30×10^{-6} was fairly close to a plot of ship-hit P_i equal to 1×10^{-5} used operationally at the WR. It may be worthwhile to conduct additional studies to assess factors not included in the analyses described here, such as the possibility that a single big hit could cause multiple casualties by completely destroying a ship or impeding crew escape. Also, comparisons based on practices at the WR do not directly apply to the ER.

Unlike RCC Standard 321-97 (RCC, 1997a), EWR 127-1 does not specify a risk standard for non-mission-essential aircraft. At both the ER and WR, several kinds of special-use airspace zones are activated to keep public aircraft totally away from hazardous operations. Restricted areas over the launch areas were established by FAA rule-making many years ago and are sized to cover a number of different sites and operations. These areas are plotted on standard aeronautical charts and are purposely large enough to create a buffer zone to compensate for the high speed of aircraft. Warning areas extend from the coast to as much as 150 miles offshore and usually take in the extremes of the ILLs (impact limit lines). The ranges also use stationary altitude reservation areas to warn aviators away from ocean areas where expended booster stages are expected to hit. These areas are also conservatively sized and include buffer zones of five miles or more. No other warnings to commercial traffic are invoked outside of these areas, and air lanes around the world continue to operate normally below the flight paths of space and missile launches.

At the WR, hazard areas inside the warning areas are defined using an aircraft P_i of 1×10^{-8}. Mission-essential aircraft are not normally allowed in hazard areas during launches. Exceptions have been made to allow mission-essential aircraft to operate in areas with a P_i up to 1×10^{-6}. In any case, air traffic is not a problem at the WR, which has not had a launch hold or scrub because of an aircraft intruder within the memory of current staff.

At the ER, P_i is used to manage risk for mission-essential aircraft, which may be allowed to operate in areas where P_i is less than 1×10^{-6}. Aircraft intrusions, generally by small private aircraft, are not uncommon. Air surveillance usually detects intruders even before they enter a restricted zone; as described later in this chapter, they are monitored during the countdown, and efforts are made to communicate with the aircraft and warn them away from the area.

Prior to each launch, the flying public is warned to stay clear of the restricted areas for that launch. Despite the large size of the standardized restricted areas, they do not seem to unduly hamper civil or commercial air operations, and using predefined areas that can be permanently plotted on aeronautical charts is much more practical than disseminating precise boundaries of warning areas for each launch using latitude and longitude coordinates. However, the use of standardized areas is not an efficient tool for determining whether a launch hold is needed. Launch hold decisions should be based on the position of intruder aircraft relative to the areas of actual hazard, especially at the ER where aircraft intrusions are more often a problem.

P_i should be calculated differently for aircraft than for ships because even small pieces of debris can endanger aircraft. For example, cylindrical steel or aluminum fragments weighing as little as 2 or 3.5 grams, respectively, can penetrate an aircraft windshield (RCC, 1997b). Such small particles are not normally taken into account in launch risk analyses. Nevertheless, an aircraft P_i limit of 1×10^{-6}, if calculated to account for very small pieces of debris, appears to be consistent with the individual ship-hit risk criteria (P_i of 1×10^{-5}) and the collective risk standard (E_c of 30×10^{-6}).

Each launch vehicle has a "commit sequence" that typically starts one to five minutes before engine ignition. Instituting a hold after the commit sequence starts can be dangerous and costly. Thus, each hazard area should be surrounded by a buffer zone large enough to ensure that aircraft will not enter the hazard area until the launch vehicle has cleared the area even if an aircraft turns toward the hazard area at the beginning of the launch commit cycle. These large buffer zones are not necessary for aircraft flying under the direction of air traffic controllers in airways outside the hazard areas.

Finding 5-1. A limit of 1×10^{-5} for individual ship-hit probability, P_i, is reasonable and consistent with an E_c of 30×10^{-6}. However, the use of *collective* risk in the Eastern Range ship exclusion process is not consistent with either the corresponding Western Range process or accepted guidelines for the evacuation of hazard areas, which are both based on *individual* risk. Aircraft avoidance criterion are not specified by EWR 127-1, are applied differently at the Eastern and Western Ranges, and are not supported by analyses showing that they are consistent with other range safety criteria.

Primary Recommendation on Risk Standards for Aircraft and Ships. AFSPC should apply the individual ship-hit criterion, P_i, of 1×10^{-5} to the ship exclusion process at the Eastern Range in the same way it is used at the Western Range. EWR 127-1 should be modified to specify an aircraft-hit P_i limit of 1×10^{-6} (properly calculated to include the probability of impact for very small pieces of debris). Prior to each launch, the range should establish aircraft hazard areas (based on the aircraft P_i) and buffer zones (for uncontrolled aircraft in the vicinity of the hazard area). Launches should be allowed to proceed as long as no intruder aircraft are in the hazard area or buffer zone.

Recommendation 5-1. AFSPC should determine maximum-acceptable blast overpressure limits and apply these limits to ship-hit calculations at both the Western and Eastern Ranges.

Guidelines and Procedures for Marine Traffic

Risks to marine traffic in and approaching a launch area are determined by flight analysis personnel using ship-hit probability contours and toxic hazard zones. Safety analyses include risks from both accidents and normal operations, including the jettison of expended stages and the areas and altitudes where toxic hazards may exist. Graphic plots are used to determine the areas that must be cleared of boats and ships (and trains, at the WR). At the ER, multiple contours are provided to show the zones that pertain to various numbers of ships or boats that are allowed within a grid area. For example, a launch would not be allowed to proceed if a boat were located inside the contour in Figure 5-1 labeled "1 boat $\times 10^{-5}$" or if two boats were located inside the contour labeled "2 boats $\times 10^{-5}$."

Extending downrange from the coastline, the launch corridor is divided into three separate areas: from the coast to three miles offshore (national waters), from 3 to 12 miles (federal waters), and beyond 12 miles (international waters). Different rules of operation apply in each of these areas with implications for surveillance, interdiction, and enforcement. Enforcement action against intruders may not be practical beyond the 12-mile limit.

General requirements for launch area surveillance and control at the ER and WR are included in EWR 127-1. Section 7.9 directs the use of surveillance aircraft and ground-based surveillance radar and requires advance public notification of specific hazard areas and times using Notices to Mariners and marine radio broadcasts. At the ER, Section 7.9 also requires the use of on-shore warning signals, lights, and signs in the vicinity of hazard areas and in channels leading to the ocean.

Guidelines for surveillance and interdiction to prevent and clear surface incursions also address the use of ground-based military and FAA radars, U.S. Coast Guard vessels (at the ER), U.S. Navy ships (at the WR), helicopters, fixed-wing aircraft, voice communications systems, and an automated

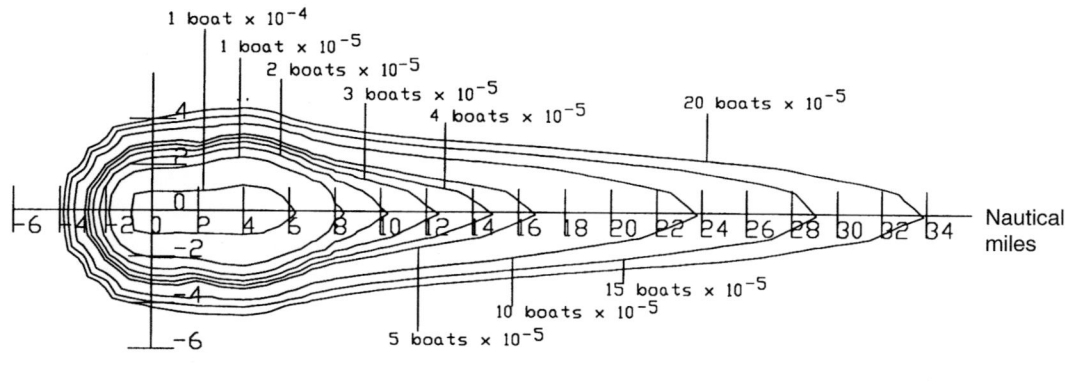

FIGURE 5-1 Samples of multiple boat-hit contours. Source: RTI, 1997.

train surveillance system (at the WR). However, many of the ships and aircraft that the ranges rely on for these services are also assigned other duties by the Coast Guard and Navy and are not always available to support launch operations.

Guidelines and Procedures for Aircraft

The safety of aircraft is ensured through the use of special-use airspace and restricted, warning, and hazard areas. The general requirements described above for launch area surveillance and control of surface traffic apply to aircraft as well. EWR 127-1 also requires advance dissemination of Notices to Airmen (NOTAMs) via the FAA to inform civil and military pilots of the boundaries for hazardous areas and the times they are active.

Nearby military radars and the FAA's air traffic control (ATC) surveillance radars are key sources of information for surveillance and clearing of launch hazard areas at both ranges. Principal sites at the ER that actively support launches include the radar approach control facility at Patrick Air Force Base and FAA radars at the Miami and Jacksonville air route traffic control centers. These radars provide aircraft surveillance within a 50-nautical mile radius of the launch site. At the WR, similar radars at Vandenberg Air Force Base and the FAA/Air Force Joint Surveillance System provide comparable air surveillance to a range of 250 nautical miles.

Communications and coordination guidelines and procedures at both ranges support effective coordination among FAA air traffic controllers and Air Force launch support staff. Air traffic controllers use radars and associated communication networks to communicate with civil and military aircraft under their control and keep them clear of hazardous airspace during launch operations. However, general aviation aircraft in the vicinity of the ranges typically fly under visual flight rules (VFR) and operate independently of ATC. Although VFR aircraft are usually equipped with operating transponders and appear on ATC radar displays, they are not uniquely identifiable, and their pilots normally are not in continuous communication with ATC. VFR traffic density at the ER is much greater and presents more of a challenge than at the WR.

Oil Rigs

Four manned oil platforms lie offshore from the WR facility within some of the launch risk corridors. Current guidelines direct that a platform be totally evacuated if E_c (the cumulative risk to personnel during launch) exceeds the criterion for mission-essential personnel (300×10^{-6}), but this has never occurred within the memory of range safety or oil industry personnel contacted by the committee. If E_c is less than 300×10^{-6}, but P_i is greater than 1×10^{-5}, unnecessary personnel are evacuated, and those who remain take shelter inside the structure of the platform during the launch period. Oil rig operators are routinely advised 10 days prior to a launch that may require personnel to be evacuated or sheltered.

Total evacuation of an oil platform could cost about $15,000 in direct expenses, plus as much as $300,000 in lost production for a total shutdown. The evacuation process also involves inherent risks to equipment and personnel (e.g., the risk associated with a helicopter flight to and from the platform). Sheltering of personnel, however, may result in only 20 to 30 minutes of lost production and little or no additional risk.

In 1988 oil platform operators examined the WR's analytical tools and the risk levels that pertain to oil platforms.

Since that time the WR has shared risk information with operators on a regular, case-by-case basis and allowed them to determine which protective measures should be taken. Reportedly, under this shared decision-making arrangement, partial evacuations are rare (no more than one per year). However, the ownership of the platforms may be changing. If so, it would be prudent for WR managers to contact the new owners and try to preserve the current, mutually beneficial partnership. In view of the cooperative arrangement between the WR and oil platform operators and the infrequent need for even partial evacuations, the committee concluded that current oil platform launch notification and evacuation criteria and procedures, as applied by the WR, are acceptable.

PLANNED IMPROVEMENTS AND ADDITIONAL RECOMMENDATIONS

Current assets to accomplish the sea surveillance and interdiction mission at the ER include a U.S. Coast Guard vessel that is subject to diversion for other search and rescue missions, a tower-mounted marine radar with a limited 20-mile range capability, and two U.S. Air Force helicopters. The helicopters are relatively slow (120 knots) and operate at low altitude (500 feet), which provides the crew with a limited field of view. The hazard area at the ER encompasses a large ocean area and frequently contains many fishing boats that routinely operate in areas with the highest hit probability. The clearing process typically involves approaching each intruder and asking the crew to move using marine VHF radio or a streamer with a written message that is dropped onto the boat. Because of the long time required for helicopters to sweep their assigned areas and the high traffic density, search crews sometimes find that an intruder has unexpectedly entered a cleared area between sweeps. As a result, it is sometimes difficult to clear hazardous launch areas and keep them clear.

Air surveillance radar data from several sites is fed to the ROCC for use in detecting and tracking aircraft in and near the launch hazard areas. The current arrangement relies on outdated radars and offers limited data storage and automatic tracking capability. As a result, considerable time and manpower are needed to integrate and interpret this information before it can be acted upon. The committee believes radar data from current ATC and military air surveillance radars may not be adequate under maximum traffic loading to provide an integrated display of aircraft position and tracking information for timely and efficient air surveillance and control purposes, especially at the ER.

Both ranges have developed improvement plans that include new surface radars and replacing military aircraft with commercial aircraft carrying improved sensor, navigation, and communications equipment. Commercial sensor technology and miniaturization have progressed to the point that military surveillance systems are no longer required. The ER estimates that the operating costs of an improved system would be about half the cost of current military helicopters, but it has not yet shifted to such a system. The committee agrees that suitably equipped, commercially operated, fixed-wing aircraft would cut costs and provide a more stable sensor platform with greater mobility, more extensive surveillance of hazard areas, and higher resolution video recording.

The Cape Canaveral Range Surveillance System (CRaSS), proposed by the ER, could also enhance air surveillance and control capabilities and possibly contribute to surface surveillance. The CRaSS system would consist of a mosaic radar display system capable of processing, storing, and displaying data from as many as 16 separate radar sources. CRaSS would greatly increase the ER's radar surveillance area and automatically track numerous targets.

The WR already is using a limited version of CRaSS. The WR also has plans for a faster and more accurate "digital rail display" for surveillance of trains transiting the range. However, because of funding and other considerations, the schedule for completing these improvements is uncertain.

Increases in launch area marine and air traffic and more frequent space launches are expected to increase the number of boat and aircraft intruders, especially at the ER. Present surveillance and interdiction capabilities at the ER and WR cannot always detect and clear incursions in time to avoid launch holds and scrubs. The committee believes that immediate improvements are warranted.

Finding 5-2. Detecting marine and aircraft intruders earlier and shortening the time required to clear them from the launch area would reduce disruptions, costs, and risks associated with launch holds and scrubs, especially at the Eastern Range where intruders are more of a problem.

Recommendation 5-2. AFSPC should expeditiously improve range surveillance and interdiction capabilities, as follows:

- Use commercial aircraft equipped with suitable surveillance, navigation, communications, and image recording systems in place of military aircraft.
- Implement the proposed Cape Canaveral Range Surveillance System (CRaSS) for surveillance and clearing of aircraft intruders at the Eastern Range.

Notices to Mariners indicating restricted areas to be avoided and the times they are active are issued before launches. Similar voice alerts are broadcast periodically over marine channels. However, these warnings are not easily accessible to many members of the pleasure-boating community, and, as a result, they are only partly effective. Public awareness and community support could be improved by more effective dissemination of information regarding hazardous launch areas. Options include public service and paid announcements on radio and television, local newspapers, the Internet, town meetings, and focused notices at marinas and commercial fishing ports. The committee believes that an aggressive, concerted education and publicity effort would greatly increase public awareness of launch hazards and reduce the incidence of both surface and airborne incursions.

NOTAMs provide similar warnings to the aviation community. Although NOTAMs are readily accessible to the pilots of most military and commercial flights, they may not be as available to general aviation pilots operating out of small or uncontrolled airports (i.e., airports without air traffic controllers). Many of these pilots do not file flight plans and do not communicate frequently with the FAA's ATC facilities. As a result, NOTAMs are only partly effective in preventing incursions by this segment of the aviation community. A more effective notification process would benefit both the general aviation community and the ranges.

Enforcement action against marine intruders inside the 12-mile limit can be taken under Title 33, Part 334 of the Code of Federal Regulations. Pilots licensed by the FAA also can be sanctioned. However, enforcement actions against intruders at either the WR or ER are rare. In contrast, at the launch site in Kourou, French Guyana, government authorities are very aggressive in taking action against intruders.

Commercial fishing vessels have substantial financial incentives to continue operations in active hazard areas, and pleasure boaters may have a strong desire to obtain a close vantage point for viewing a launch. They may also be reluctant to interrupt their activities to leave proscribed areas. The frequent incursions of pleasure boaters also demonstrates that the perceived level of personal risk is not an effective deterrent. Finally, intruders rarely face direct penalties, or even censure, for noncompliance.

The situation is much the same with general aviation pilots. Having the equivalent of "front row seats" to a spectacular and dramatic, if not historic, event is understandably appealing to many people. Both the Atlantic and Pacific coastal areas of the United States offer scenic views for pleasure fliers. Because many general aviation pilots don't understand the risk that a launch accident could pose to their aircraft, it is not surprising that intruders are sometimes a problem.

Finding 5-3. Current guidelines and procedures for notifying operators of general aviation aircraft and small boats of active launch hazard areas do not prevent incursions, especially at the Eastern Range.

Recommendation 5-3. AFSPC should improve the launch communications and notification process, as follows:

- Make greater use of public media, such as newspapers, radio and television broadcasts, the Internet, notices at public marinas and general aviation airports, and aviation and marine weather broadcasts.
- Modify signs, lights, and other warning devices at marinas and along the coast, as necessary.
- Inform the public on the extent of safe viewing areas to discourage operators of small boats and aircraft from encroaching on hazard areas.

Although an aggressive notification and education process directed at the flying and boating public is expected to contribute significantly to increasing community support and reducing the number of incursions, additional action is needed to address the intruder problem.

Recommendation 5-4. In combination with efforts to improve surveillance and interdiction capabilities and the public notification process, AFSPC should aggressively enforce restrictions against intruders at both ranges to encourage compliance with launch notifications. In cooperation with the U.S. Coast Guard, the Federal Aviation Administration, the U.S. Attorney's Office, and other regulatory and law enforcement agencies, AFSPC should initiate administrative and regulatory changes to facilitate enforcement action against intruders who were afforded ample, timely launch notifications.

REFERENCES

EWR 127-1 (Eastern and Western Range Safety Requirements). 1997. Available on line at: *http://www.pafb.af.mil/45sw/rangesafety/ewr97.htm* January 20, 2000.

RCC (Range Commanders Council). 1997a. Common Risk Criteria for National Test Ranges: Inert Debris. AD-A324356. RCC Standard 321-97. February 1997. Available on line at: *http://www.jcte.jcs.mil/RCC/oldoc.htm* January 28, 2000.

RCC. 1997b. Supplement to Common Risk Criteria for National Test Ranges: Inert Debris. AD-A324955. February 1997. Available on line at: *http://www.jcte.jcs.mil/RCC/oldoc.htm* January 28, 2000.

RTI (Research Triangle Institute). 1997. Flight Control and Analysis General Reference Handbook. RTI/6762/03-02F. Prepared for 30th and 45th Space Wings, April 24, 1997. Patrick Air Force Base, Fla.: 45th Space Wing.

Appendix A

Findings and Recommendations

CHAPTER 2

Background

Finding 2-1. Range safety personnel and procedures have well protected people and property. In the history of the U.S. space program, no members of the general public or launch site workers have been killed or seriously injured during a launch accident.

CHAPTER 3

Risk Management Approaches to Safety

Primary Recommendation on EWR 127-1. AFSPC should simplify EWR 127-1 so that all requirements are performance based and consistent with both established risk standards for space launch (e.g., E_c of 30×10^{-6}) and objective industry standards. The process of revising EWR 127-1 should include the following steps:

- Eliminate requirements that cannot be validated.
- Remove all design solutions from EWR 127-1.
- Establish a range user's handbook or other controlled document to capture lessons learned and design solutions recognized by the ranges as acceptable means of compliance. (Requirements should be retained in EWR 127-1.)
- Form a joint government/industry team to establish procedures for periodically updating EWR 127-1 and ensuring that future requirements are performance based.
- Converge the modeling and analysis approaches, tools, assumptions, and operational procedures used at the Western and Eastern Ranges.

Finding 3-1. AFSPC has transferred responsibility to AFMC for development, developmental testing and evaluation, and sustaining engineering of range safety ground systems. Organizational responsibilities for many other range safety processes and procedures, however, are inconsistent with the current memorandum of agreement between AFSPC and AFMC on spacelift roles and responsibilities. In addition to the operational workforce, each AFSPC range safety office also has an engineering workforce that establishes flight safety system design and testing requirements and certifies that flight safety systems meet safety requirements at the component, subsystem, and system levels. These acquisition-like functions overlap the responsibilities of AFMC.

Finding 3-2. The complete transfer of range safety development, developmental testing and evaluation, and sustaining engineering to AFMC would, if properly implemented, increase efficiency and reduce costs without compromising safety by eliminating overlapping responsibilities between the ranges and AFMC, by minimizing differences in range safety policies and procedures applicable to the Western and Eastern Ranges, and by enabling users to deal with a single office when seeking approval to use new or modified systems on both ranges.

Primary Recommendation on Roles and Responsibilities. The Air Force should fully implement the memorandum of agreement between AFSPC and AFMC on spacelift roles and responsibilities. This would consolidate within AFMC the acquisition-like functions related to safety that are now performed by AFSPC organizations at the Eastern and Western Ranges. These functions include developmental testing and evaluation, sustaining engineering, and certifying that system designs meet safety requirements. To manage the safety aspects of the acquisition-like functions specified in the memorandum of agreement, AFMC should establish an independent safety office. Operational responsibilities, such as generating safety requirements, operational testing and evaluation, and all prelaunch and launch safety operational functions, would be retained by AFSPC.

Recommendation 3-1. AFSPC should issue an Air Force Instruction addressing the certification of flight safety systems for commercial, civil, and military launches at the Western or Eastern Range. The instruction should include a description of interfaces among responsible organizations, such as AFSPC, AFMC, FAA, NASA, and commercial contractors.

Finding 3-3. A collective risk standard (i.e., a casualty expectation, or E_c) of 30×10^{-6} per launch for members of the general public is consistent with the risk standards of many other fields in which the public is involuntarily exposed to risk, both domestically and internationally.

Primary Recommendation on Risk Management. AFSPC should define objective, consistent risk standards (e.g., casualty expectation, E_c, of 30×10^{-6} and individual risk, P_c, of 1×10^{-6}) and use them as the basis for range safety decisions. Safety procedures based on risk avoidance should be replaced with procedures consistent with the risk management philosophy specified by EWR 127-1. Destruct lines and flight termination system requirements should be defined and implemented in a way that is directly traceable to accepted risk standards.

Finding 3-4. At the Eastern Range, the downrange location of gates and destruct lines and current requirements for downrange coverage by flight termination, telemetry, and tracking systems are not directly related to accepted risk standards (e.g., E_c of 30×10^{-6} or P_c of 1×10^{-6}) but to a risk-avoidance policy that discourages the overflight of inhabited landmasses whenever possible. The Western Range implements this policy by constraining the azimuth of orbital launches.

Finding 3-5. Moving the Africa gates uprange has the potential to reduce the cost of safety-related downrange assets, decrease the complexity of range safety operations, and reduce launch holds and delays. Moving the Africa gates to within the reach of uprange flight termination, telemetry, and tracking systems is not likely to increase E_c significantly or violate established limits. No known international agreements would preclude moving the gates. Thus, in terms of range safety there is no clear justification for retaining downrange assets at Antigua and Ascension. It may also be feasible to move other gates uprange and further reduce the need for downrange facilities.

Primary Recommendation on Africa Gates. While other requirements may exist, from the perspective of launch range safety the Air Force should move the Africa gates to within the limits of uprange flight termination and tracking systems; eliminate the use of assets in Antigua and Ascension for range safety support; and conduct a detailed technical assessment to validate the feasibility of moving other gates uprange. If other requirements for downrange tracking exist, AFSPC should validate those requirements and reexamine this recommendation in light of the additional requirements.

Recommendation 3-2. AFSPC should identify and correct unwarranted conservatism in analytical models and verify that modeling and analytical methods are properly implemented. Periodic, independent reviews should be conducted to ensure that the level of modeling detail is appropriate given the accuracy of model inputs and assumptions.

Finding 3-6. The overall modeling and analysis approaches at the Eastern and Western Ranges are similar, but there are some significant differences in analytical tools, assumptions, and operational procedures. These include differences in analysis software packages, methods of defining ship exclusion zones, and displays for monitoring the launch vehicle trajectory. The differences may increase costs because of overlap or duplication of effort in developing models, software, and hardware for the two ranges.

CHAPTER 4

Flight Safety Requirements

Recommendation 4-1. As a matter of good engineering practice, the requirement for two independent sources of tracking data should be retained, and the accuracy of telemetered inertial guidance data should be verified after launch. AFSPC should clarify EWR 127-1 to specify that telemetered inertial guidance data can serve as one of the two sources of tracking data.

Finding 4-1. For space launches, an onboard GPS receiver tracking system would be more versatile and have lower total life-cycle costs than GPS translator or radar tracking systems.

Finding 4-2. Real-time GPS tracking systems have an overall cost and performance advantage over the single-object radar network that has been the workhorse on both the Eastern and Western Ranges for many years. Implementation of a GPS tracking system would increase users' recurring and nonrecurring costs in the short term, but it would benefit users in the long term by increasing operational flexibility. A GPS tracking system would also yield long-term costs savings for the ranges.

Primary Recommendation on GPS Receivers. AFSPC should deploy a GPS receiver tracking system as the baseline range tracking system for space launch vehicles. The transition to GPS-based tracking should be completed as rapidly as feasible.

Finding 4-3. Upgrades to onboard tracking systems currently in use and to new systems, such as GPS receivers, are relatively costly for individual users. Each user currently must develop or acquire hardware, prove that it meets safety requirements, demonstrate its compatibility with range support equipment, provide for qualification and acceptance testing, and support confidence checks in the final countdown.

Recommendation 4-2. AFSPC should form a range-industry team to define performance requirements and technical specifications for the onboard elements of a GPS receiver tracking system, including cost, weight, size, and power limitations, and to establish user requirements during the transition from radar to GPS-based tracking systems. A cost-shared government/industry project should be established for the development and qualification testing of common end-user equipment. Range users should pay for the recurring costs of onboard hardware.

Finding 4-4. With the incorporation of onboard GPS receivers, semiautonomous and fully autonomous flight termination systems would become technically feasible. These systems might substantially reduce range support costs, but additional research and testing is needed to resolve outstanding issues and quantify the likely benefits.

CHAPTER 5

Incursions

Finding 5-1. A limit of 1×10^{-5} for individual ship-hit probability, P_i, is reasonable and consistent with an E_c of 30×10^{-6}. However, the use of *collective* risk in the Eastern Range ship exclusion process is not consistent with either the corresponding Western Range process or accepted guidelines for the evacuation of hazard areas, which are both based on *individual* risk. Aircraft avoidance criterion are not specified by EWR 127-1, are applied differently at the Eastern and Western Ranges, and are not supported by analyses showing that they are consistent with other range safety criteria.

Primary Recommendation on Risk Standards for Aircraft and Ships. AFSPC should apply the individual ship-hit criterion, P_i, of 1×10^{-6} to the ship exclusion process at the Eastern Range in the same way it is used at the Western Range. EWR 127-1 should be modified to specify an aircraft-hit P_i limit of 1×10^{-6} (properly calculated to include the probability of impact for very small pieces of debris). Prior to each launch, the range should establish aircraft hazard areas (based on the aircraft P_i) and buffer zones (for uncontrolled aircraft in the vicinity of the hazard area). Launches should be allowed to proceed as long as no intruder aircraft are in the hazard area or buffer zone.

Recommendation 5-1. AFSPC should determine maximum-acceptable blast overpressure limits and apply these limits to ship-hit calculations at both the Western and Eastern Ranges.

Finding 5-2. Detecting marine and aircraft intruders earlier and shortening the time required to clear them from the launch area would reduce disruptions, costs, and risks associated with launch holds and scrubs, especially at the Eastern Range where intruders are more of a problem.

Recommendation 5-2. AFSPC should expeditiously improve range surveillance and interdiction capabilities, as follows:

- Use commercial aircraft equipped with suitable surveillance, navigation, communications, and image recording systems in place of military aircraft.
- Implement the proposed Cape Canaveral Range Surveillance System (CRaSS) for surveillance and clearing of aircraft intruders at the Eastern Range.

Finding 5-3. Current guidelines and procedures for notifying operators of general aviation aircraft and small boats of active launch hazard areas do not prevent incursions, especially at the Eastern Range.

Recommendation 5-3. AFSPC should improve the launch communications and notification process, as follows:

- Make greater use of public media, such as newspapers, radio and television broadcasts, the Internet, notices at public marinas and general aviation airports, and aviation and marine weather broadcasts.
- Modify signs, lights, and other warning devices at marinas and along the coast, as necessary.
- Inform the public on the extent of safe viewing areas to discourage operators of small boats and aircraft from encroaching on hazard areas.

Recommendation 5-4. In combination with efforts to improve surveillance and interdiction capabilities and the public notification process, AFSPC should aggressively enforce restrictions against intruders at both ranges to encourage compliance with launch notifications. In cooperation with the U.S. Coast Guard, the Federal Aviation Administration, the U.S. Attorney's Office, and other regulatory and law enforcement agencies, AFSPC should initiate administrative and regulatory changes to facilitate enforcement action against intruders who were afforded ample, timely launch notifications.

Appendix B

Biographies of Committee Members

ROBERT E. WHITEHEAD (chairman) entered government service in 1971 after receiving undergraduate and graduate degrees in engineering mechanics from Virginia Polytechnic Institute and State University and completing one year of postdoctoral study at the National Aeronautics and Space Administration (NASA) Ames Research Center. Dr. Whitehead began his career with the Department of the Navy as a research engineer in the Aviation Department of the David Taylor Naval Ship Research and Development Center at Carderock, Maryland. He transferred to the Office of Naval Research (ONR) in 1976 as a scientific officer in applied aerodynamics. For the next 13 years, Dr. Whitehead held a number of positions at ONR, including director, Mechanics Division, from 1986 to 1989, when he transferred to NASA Headquarters, as the assistant director for aeronautics (rotorcraft). He held a variety of other positions before becoming the associate administrator for aeronautics in 1995 and associate administrator for aeronautics and space transportation technology in 1997. As associate administrator, Dr. Whitehead led a Research and Technology Enterprise of more than 6,000 civil servants and a similar number of contractors at four research centers with an annual budget of approximately $1.5 billion. Dr Whitehead retired from federal service in December 1997.

W. GAINEY BEST II has been employed by Lockheed Martin Astronautics since 1994. In 1997, Mr. Best became director of the Evolved Expendable Launch Vehicle (EELV) Program, assuming responsibility for cost, schedule, technical baseline, and performance. Previously, he led the EELV Mission Integration Team. Before that, he was responsible to the director of the Titan Centaur Program for independent assessments of technical readiness for building, launch processing, and launching of each Titan Centaur vehicle. Mr. Best entered the U.S. Air Force in 1968, where he was the director of West Coast operations for the National Reconnaissance Office (NRO). He directed daily operations, including worldwide elements involved in the design, acquisition, launch, deployment, and orbital operation of satellites. He was also the program manager and the deputy program director for the Titan IV launch vehicle program. During his career in the Air Force, Mr. Best was the Air Force lead for the investigation of three launch failures. His experience includes many operational, satellite, and program management assignments. He has earned degrees in industrial management and mechanical engineering.

JOHN L. BYRON works in strategic planning, process engineering, and corporate planning for Johnson Controls of Cape Canaveral, Florida. For the past five years, he has been vice chairman of the Florida Space Business Roundtable. He is also a member of on the Board of Directors of the newly formed Florida Space Research Institute. Mr. Byron retired from the Navy in 1993 after more than 37 years of continuous active duty. During that time he commanded the Naval Ordnance Test Unit (NOTU) at Cape Canaveral, where he was director of Navy tests for the Eastern Range and supervised the launch of 52 Trident missiles from submerged submarines. Mr. Byron has a B.S. in physical oceanography from the University of Washington, and he is a graduate of The National War College.

BENJAMIN A. COSGROVE is a retired senior vice president of Boeing Commercial Airplane Group. His 43-year career as a structural engineer began at Boeing in 1949 on the B47 and B52 bombers. He was involved in the design and analysis of every Boeing commercial airplane from the 707 through the 777. Mr. Cosgrove was the chief design engineer of the 767, became vice president of engineering and flight testing in 1985, and was promoted to senior vice president in 1989. The National Aeronautic Association of Washington, D.C., has awarded him the Wright Brothers Memorial Trophy for his lifetime contributions to commercial aviation safety and technical achievement. He is a member of the National Academy of Engineering and received an honorary doctorate of engineering from the University of

Notre Dame. Mr. Cosgrove is also a member of the NASA Advisory Council's Task Force on the Shuttle-Mir Rendezvous and Docking Missions and the Task Force on International Space Station Operational Readiness, which are chaired Lt Gen Thomas Stafford, USAF (retired).

JAMES W. DANAHER is the retired chief of the Operational Factors Division of the Office of Aviation Safety at the National Transportation Safety Board (NTSB) in Washington, D.C. He has more than 35 years of experience in human factors and safety, in both industry and government. Since he joined the NTSB in 1970, Mr. Danaher has served in various supervisory and managerial positions, with special emphasis on human performance issues in flight operations and air traffic control. He has participated in the on-scene phase of numerous accident investigations, in associated public hearings, and in the development of NTSB recommendations for the prevention of future accidents. He is a former naval aviator and holds a commercial pilot's license with single-engine, multi-engine, and instrument ratings. He has an M.S. degree in experimental psychology from Ohio State University and is a graduate of the Federal Executive Institute. Mr. Danaher has represented the NTSB at numerous safety meetings, symposia, and seminars, is the author or co-author of numerous publications, and served on the National Research Council's Panel on Human Factors in Air Traffic Control Automation.

KINGSTON A. GEORGE is a retired chief engineer for safety from the 30th Space Wing, Vandenberg Air Force Base, California. In 1959, he completed a unique five-year combined degree program at Ohio State University, graduating with a B.S. in engineering physics and an M.S. in nuclear physics. After two years as a researcher at the cyclotron laboratory at Ohio State, he joined the Operations Analysis Office under the Air Force at Vandenberg Air Force Base in 1961, where he was engaged in evaluating the capability and test design for ballistic missiles. He then moved to the newly formed Air Force Western Test Range in 1965, where his accomplishments include defining controlled areas during launch; establishing telescopic camera sites for engineering data; and developing improved data processing methods for real-time display and flight control. He was a member of the Range Commanders Council and chairman of the Executive Committee for one term. He also chaired a tri-service study on the use of GPS for launch ranges that culminated in a major project at Eglin Air Force Base funded by the U.S. Department of Defense to design and build GPS-based test range instrumentation. A member of Tau Beta Pi, Sigma Pi Sigma, and the American Physical Society, Mr. George was honored in 1989 as the Air Force Association's Senior Manager of the Year. He is currently a senior consultant in the aerospace industry, primarily concerned with issues of real-time tracking and flight safety for space launch programs.

BILL HAWLEY has been employed by Hughes Space and Communications since 1978, where he currently is manager of Launch Systems Engineering and Operations. His responsibilities include all launch vehicle integration and engineering for Hughes spacecraft, system safety, and launch operations. Previously, as department manager of propulsion engineering, Mr. Hawley was responsible for propulsion system design and component development, including procurement, testing, and propellant loading operations. He has also been project manager and department manager for Spacecraft Structures and Integration; head of the Mechanical Ground Support Equipment Section; and structural designer for meteorological satellites. Mr. Hawley also has worked as a design engineer for Rockwell International. He received a B.S. degree in aerospace engineering from the California State Polytechnic University.

JAMES K. KUCHAR is an assistant professor of aeronautics and astronautics at the Massachusetts Institute of Technology (MIT), where he has been on the faculty since 1995. His research interests are focused on safety-critical decision aiding and alerting systems, risk assessment, advanced cockpit displays, air traffic control, and flight simulation. He has performed several risk assessment studies of instrument approaches to closely spaced parallel runways for NASA, the Federal Aviation Administration, and Draper Laboratory. These studies developed methods to estimate risk during parallel approaches and provided system design guidelines to balance risks against landing capacity. He has also investigated policy issues related to air traffic and space launch operations. For his work on alerting systems, Dr. Kuchar was awarded the RTCA William E. Jackson Award and the Council for University Transportation Centers' Wootan/Pikarsky Award in 1995. He received his S.B., S.M., and Ph.D. in aeronautics and astronautics from MIT, where his work focused on terrain displays for transport aircraft. Dr. Kuchar is a member of the American Institute of Aeronautics and Astronautics, and he is a private pilot. Dr. Kuchar currently teaches courses on estimation, numerical methods, flight simulation, and decision aiding and alerting systems.

JOYCE A. McDEVITT is a program manager with Futron Corporation, Washington, D.C., where she provides range safety and system safety support to government and commercial clients. She is currently supporting the Commercial Space Transportation Licensing and Safety Division of the Federal Aviation Administration. Ms. McDevitt has more than 30 years of experience in safety involving space, aeronautical, facility, and weapons systems, including propellant, explosive, and chemical processes. She has developed and managed safety programs, hazard analyses, safety risk assessments, safety policies and procedures, investigations of mishaps, and safety training. She retired from the federal government in 1987 after working for NASA Headquarters, Air Force Systems Command, and the Naval Ordnance

Station. Ms. McDevitt received a B.S. in chemical engineering from the University of New Hampshire and an M.S. in engineering from Catholic University. She is a registered professional engineer in safety engineering and an active member of the System Safety Society.

JOSEPH MELTZER, currently a system engineering and planning consultant for Space and Missile Systems, retired in September 1997 as corporate chief engineer of the Aerospace Corporation. In this position, he was responsible for systems engineering policies and practices, cross-program engineering, integrated weapons systems management, and product development systems, as well as acquisition process improvements, such as risk management, safety, security systems, readiness reviews, failure analysis, and launch certification. The Aerospace Corporation is a nonprofit, federally funded research and development center that provides general systems engineering support to the Air Force and other government space-related programs. Since joining Aerospace in 1963, Dr. Meltzer has held positions of increasing responsibility, including director of spacecraft programs and general manager of the Eastern Technical Division, which was located in Washington, D.C. Prior to joining the Aerospace Corporation, he worked on missile systems for Hughes Aircraft Company, on missile systems and reentry vehicles for Lockheed Aircraft Company, and on spacecraft propulsion systems for Giannini Scientific Corporation. Dr. Meltzer earned his B.S., M.S., and Ph.D. in engineering from the University of California, Los Angeles.

JIMMEY MORRELL has more than 30 years of experience in a wide range of technical, management, and administrative activities. After retiring from the U.S. Air Force, Maj Gen Morrell became senior vice president and director of the Decision Technologies Division at GRC International, Inc., where he managed Air Force space and classified business activities. As an Air Force officer, Maj Gen Morrell served as a senior policy analyst at the White House Office of Science and Technology Policy, assistant chief of staff of the Air Force Air University, commander of an Air Force satellite control wing, and commander of the 45th Space Wing at Cape Canaveral Air Station. Maj Gen Morrell was also a congressional liaison for the Air Force Office of Space Systems and the military assistant to the Secretary of the Air Force.

NORMAN H. SCHUTZBERGER is the director of the Fluid Mechanical and Propulsion Division of TRW Components International, which engineers and supplies systems, subsystems, and components to international commercial and manned space flight system manufacturers. Mr. Schutzberger earned his B.Sc. degree in mechanical engineering from Pratt Institute and an Executive M.B.A. from the Peter F. Drucker Graduate Management Center of the Claremont Graduate University. He began his career as an engineering co-op student in optics at NASA's Goddard Space Flight Center. After graduation from Pratt Institute, he moved into an advanced spacecraft design and analysis branch at NASA, where he analyzed early designs for the Space Telescope and Earth Observing System Satellites. Subsequent positions included resident mechanical systems manager for the Delta Launch Vehicle and Space Shuttle Upper Stages, where his responsibilities included mechanical, hydraulic, and payload attachment structures and mechanisms, as well as ordnance ignition, separation, and launch range safety flight termination systems. Mr. Schutzberger has been involved in the direction, integration, and launch of 85 international, commercial, and scientific satellites. He has been an internal consultant to NASA flight programs, NASA's representative on commercial satellite failure review boards, and an invited consultant to Lloyd's for assessing satellite launch risks.

FREDERICK H. HAUCK, Aeronautics and Space Engineering Board liaison to the Committee on Space Launch Range Safety, is president and chief executive officer of AXA Space, Bethesda, Maryland. AXA Space, a member of the global AXA insurance group, specializes in underwriting the risk of launching and operating space systems. Before joining AXA Space in 1990, Mr. Hauck completed a 28-year career in the U.S. Navy as a combat pilot, test pilot, and astronaut. His last military assignment was director of Navy Space Systems in the Pentagon. During his 11 years as a NASA astronaut, he flew on three space shuttle missions, the last as commander of *Discovery* on the first space shuttle mission after the *Challenger* tragedy. Mr. Hauck received a B.S. in physics from Tufts University and an M.S. in nuclear engineering from MIT. He is a member of the Board of Trustees of Tufts University and of the American Astronautical Society and a fellow of both the Society of Experimental Test Pilots and the American Institute of Aeronautics and Astronautics. Mr. Hauck was awarded two Defense Distinguished Service Medals, the NASA Distinguished Service Medal, and the Distinguished Flying Cross. He has been a member or chair of numerous panels and advisory groups on national and international space issues.

Appendix C

Participants in Committee Meetings

The full committee met four times from April through August 1999. As part of the committee's information-gathering process, several smaller meetings were attended by one or more committee members and representatives of public and private organizations involved in space launch range safety. The open portions of the full committee meetings were attended by many individuals who were not specifically invited; the list below includes the names of participants who signed in, as well as the names of invited speakers.

ACTA, Inc.
James Baeker
Jon Collins
Fred Grimes
Karl Overbeck
Carlton Parks
Harold Reck

The Aerospace Corporation
William Butler
John Cameron
Gerald Finn
Mike Foehner
John Genovese
James Gin
Gail Johnson
Norman Keegan
Manuel Landa
Paul Mohlman
Rudy Mostajo
John Neeson
Douglas Schulthess
Bruce Simpson
Michael Spence
Paul Utecht
Joe Wambolt
Phillip Wildhagen
William Zelinsky

The Boeing Company
William Hampton
Kathleen McLaughlin

Kip Mikula
Ricardo Navarro
Rich Nieduhauser
Wayne Owens
Jack Schweikert
L. Yearsley

Cincinnati Electronics
Joseph Hermann
William Lampe

City of Cape Canaveral, Florida
Rocky Randels

Command and Control Technologies Corporation
Kevin Brown
Peter Simons

Computer Sciences Corporation
Larry Shelley

Computer Sciences Raytheon
Michael Maier
Mickey Olivier

Consultants
Bob Brewster
Paul Echerd
Richard Lee
David Richardson
Robert Stahl

DynSpace
Bob Parker

ENSCO, Inc.
Karen Haenke
Ron Ostroff
David Smith

Federal Aviation Administration
Ronald Gress

Florida Solar Energy Center
Todd Halverson
Ross McCluney

Florida Today
Malcolm Denemark
Todd Halvorson
Harvey Taffet

The Hauna Studio
Hal Hauna

Hernandez Engineering, Inc.
Maxie Peterson

Honeywell
Carlos Fernando Vales

Infoware Systems, Inc.
Bob Augustine

ITT Systems
Ken McCaniel
H. Spaulding

Ken Jongebloed, Inc.
Ken Jongebloed

Kistler Aerospace Corporation
Paul Birkeland
Jack Gregory

L3 Communications-Interstate Electronics
Lawrence Wells

Lockheed Martin
Frank Bell
Edward Butt
Thomas Hillyer
Michael Murray
Thomas Palmer
Earl Porco
Lynn Smith
Ben Ward

National Aeronautics and Space Administration
Mark Berte
John Hudiburg
Les McGonigal
Greg Oliver
Loren Shriver
Albert Sofge
Bruce Underwood

Office of Dave Weldon, Member of Congress, 15th District, Florida
Pamela Gillespie

Orbital
Frank Bellinger
Chris DeMars

Pioneer Consulting
Raymond Toomey

Raytheon Systems Company
Roger Evans

Research Triangle Institute
Gerald Bieringer
Kenneth Kaisler
Loyd Parker
Jack Parks

Santa Maria Public Airport District, California
Theodore Eckert

Sebastian Inlet Tax District, Florida
Pat Hartman

Santa Maria Times
Janene Scully

Sigmatech, Inc.
Phil Hays

Spaceport Florida Authority
Kenneth Gunn
Albert Thomas
Louis Ullian
Keith Witt

SRI International
Jonathan Brown
James Means
Gerald Shaw

SRS Technologies
Chris Komatinsky
Peter Mazur
Michael Slusher
Brian Strohman
Trase Travers

Town of Indialantic, Florida
Bob Hartman

University of Central Florida
Ross McCluney

U.S. Air Force, 30th Space Wing
Lance Adkins
Leo Aragon
Michael Cancellier
Paul Klock
Michael McCombs

U.S. Air Force, 45th Space Wing
Darren Bergan
Dan Berlinrut
William Breyer
Michael Campbell
Barry Chefer
Frank Davies
Roger Devivo
Steve Duresky
Carl Haulk
John Kinstle
Tom Palo

Paul Rosati
John Sienkiewicz
Dave Stone
Bruce Syarto
Wayne Thompson

U.S. Air Force, Aeronautical Systems Center
Cheri Hammer

U.S. Air Force Safety Center
Catherine Zeringue

U.S. Air Force Space Command
Tim Clapp
Alfred Cox
Mark Dowhan
Dave Fox
Joe Fury
Jeffrey Hill
John McConne
Joe Nemeth
M. Roney
Tim Slavenwhite
David Thompson

U.S. Air Force Space and Missile Systems Center
Les Bordelon
Loz Enas
Thomas Fitzgerald
Homer Tackett
Thierny Woods

U.S. Navy
Douglas Burnett
Jeffrey Kirchmer
Steven Landau
Javier Sanchez
George Williams
R. Williamson

WESH Television, Inc.
Cyndy Russell

TRW
Jean Daniels

Others
Richard Henry
Dave Huff
Ken Kaisler
Chuck Mertz
Harvey Taffet
Jo Townsend

Appendix D

Studies Related to Space Launch Range Safety

This appendix contains brief descriptions of 15 recently completed or ongoing studies related to U.S. space launch.

Range Integrated Product Team (IPT) Report. In the fall of 1998, Air Force Space Command (AFSPC) commissioned a study on how to make their space launch ranges more efficient and customer friendly. The study was conducted by Air Force officers (retired and active) and others from government and industry. In mid-December 1998, AFSPC presented the results of this study to the U.S. Department of Defense (DoD), U.S. commercial space companies, the National Aeronautics and Space Administration (NASA), and the Federal Aviation Administration (FAA). AFSPC then formed teams to develop implementation plans for each of the study's recommendations. Recommendations that could affect agency roles and responsibilities were remanded to Headquarters, U.S. Air Force.

Streamlining Space Launch Range Safety. One recommendation of the *Range IPT report* described above was that AFSPC sponsor "an independent technical assessment by the National Academy of Sciences of Air Force public safety methods and processes." This report is the direct result of that recommendation.

Commercial Space Opportunities Study. In December 1998, the Air Force Chief of Staff initiated a study to identify opportunities for the Air Force to take advantage of developments in the growing commercial space sector. Separate teams addressed communications, remote sensing, navigation, launch range and satellite control, and launch. Each team included representatives of the Air Force, National Reconnaissance Office, and DoD. Some teams also included representatives of NASA, the U.S. Department of Commerce, the FAA, and industry. The final report, which was completed in October 1999, concluded that the Air Force could benefit from more commercial involvement in space launch activities.

National Launch Capabilities Study. The Commercial Space Act of 1998 directed the secretary of defense to compare future space launch requirements to current capacity and address any shortfalls and funding responsibilities. The Air Force led this study, as specified in the act, and coordinated its activities with other elements of DoD, as well as the U.S. Department of Transportation, the U.S. Department of Commerce, and NASA. The report, was delivered to Congress on July 8, 1999.

A Space Roadmap for the 21st Century Aerospace Force. In November 1998, the Air Force Scientific Advisory Board completed a study that recommended changes for making the best use of space to accomplish the Air Force's operational tasks in the twenty-first century. One section of the report noted that "deteriorating facilities and an increasingly commercial launch schedule create a serious Air Force burden." The study recommended "transitioning national launch facilities to civilian operations with the Air Force as a tenant."

National Security Council/Office of Science and Technology Policy Interagency Review. On March 29, 1999, the assistant to the president for science and technology and the assistant to the president for national security affairs requested that the secretaries of defense, commerce, and transportation, the director of the Central Intelligence Agency, and the NASA administrator participate in an interagency review on the future management and use of the U.S. space launch bases and ranges, to be co-chaired by the National Security Council and the Office of Science and Technology Policy. The review assessed civil, commercial, and national security roles and responsibilities for operations, maintenance, and modernization at the space launch bases and ranges. The formal review began in April 1999, and the interim recommendations were considered in formulating the fiscal year 2001 budget. The final report was released on February 8, 2000, and is available on line (*http://www.whitehouse.gov/WH/EOP/OSTP/html/0029_6.html*).

Defense Science Board Study of Air Force-Commercial Space Launch. The fiscal year 2000 Defense Authorization Bill requires that the secretary of defense study the future of Air Force and commercial space launches and submit a report in February 2000. The Defense Science Board has been assigned to conduct the study, which is now under way.

Assessing and Achieving Customer Satisfaction at the Florida Launch Site. J.D. Powers and Associates built a customer model for the Cape Canaveral/Kennedy Space Center complex at the behest of NASA's Kennedy Space Center, the Air Force's 45th Space Wing, Enterprise Florida, Boeing, and Lockheed Martin. A report on the customer model, released on July 9, 1999, makes three comments significant to the task of the Committee on Space Launch Range Safety:

- Safety is paramount to all launch range customers.
- The customer interface with launch range management (including launch range safety offices) must be simplified.
- Safety should undergo "zero-based rethinking."

Also in 1999, NASA's Kennedy Space Center, Spaceport Florida Authority, Enterprise Florida, and the U.S. Air Force 45th Space Wing agreed to form a "spaceport management council" to manage and coordinate activities at the Kennedy Space Center and Cape Canaveral Air Station. The nature of this body and the participation of the federal agencies are still emerging.

Volpe Center Study of Florida's Role in Space. In 1999, Florida Governor Jeb Bush commissioned the Volpe Center to conduct a study of Florida's role in space and its current status. The report is pending. A preliminary executive summary of the study is now circulating in the Florida state government, and a final draft is expected early in 2000.

Aerospace Corporation Review of EWR 127-1. As a result of the *Range IPT Report*, the Air Force Space and Missile Systems Center chartered the Aerospace Corporation to characterize all requirements listed in EWR 127-1 and, if possible, determine the source or justification for each regulation. The review has been completed and results forwarded to AFSPC.

Review of EWR 127-1 by the 30th and 45th Space Wings. The Space Wings also are reviewing EWR 127-1, mindful of the recommendations that have been and are being generated by the other studies listed in this appendix. An updated version of EWR 127-1 may be issued by the summer of 2000. In addition, both Space Wings routinely review launch range safety issues as part of their normal mission, often with the engineering and analysis support of outside consulting firms.

Broad Area Review of Access to Space. The president directed the secretary of defense to report on the causes of recent failures of government space launches and to determine what actions should be taken to ensure future access to space. In response, AFSPC appointed retired Air Force Chief of Staff, General Larry Welch, to conduct a broad area review of the development, production, preparations, and launch of recent missions that ended in failure. The review included personnel from the Central Intelligence Agency, FAA, NASA, National Reconnaissance Office, and the Office of Science and Technology Policy. The resulting report, which was released in November 1999, focuses on actions the government should take to improve mission success in the future, especially with regard to the Evolved Expendable Launch Vehicle. The results of the study are available on line (*www.af.mil/lib/misc/spacebar99.htm*).

American Institute of Aeronautics and Astronautics (AIAA) Development of a National Standard for Commercial Launch Safety. The AIAA Space Launch Systems Committee conceived the idea of a commercial launch safety standard in 1992. Years of effort and many drafts later, a standard was approved on October 5, 1999, that provides guidelines for defining the safety responsibilities of ranges and users, safety requirements, and launch safety processes.

International Standards Organization Development of International Standards for Safety Requirements in Launch Site Operations. Parallel to the AIAA effort described above, the International Standards Organization has been developing an international standard, Space Systems—Launch Site Operations—Safety Requirements (ISO DIS 14620), for use wherever ISO standards are invoked. This standard is still undergoing review and development.

Department of Transportation Range Safety Standards. In accordance with its rule-making responsibilities and cognizance over the licensing of space launch operations for commercial launchers, the FAA, on behalf of the U.S. Department of Transportation, has been developing its own safety standards. Largely derived from EWR 127-1, these emerging FAA regulations have the potential to become a national standard.

Appendix E

Safety Modeling and Analysis

The primary hazards from launch accidents are associated with debris, toxic effects, and blast overpressure. Debris is created by aerodynamic forces that break up the vehicle, by explosions caused by system malfunctions, or, in many cases, as the intended result of initiating flight termination. Toxic effects may be caused by effluents from launches or catastrophic accidents. Vehicle explosions may also create blast overpressure, which can break windows and cause injuries from glass fragments miles from an accident site. Modeling of these effects is needed for launch safety.

PRELAUNCH MODELING

Nominal trajectory and expected variations from nominal. The launch customer generally provides mission data describing the nominal vehicle trajectory and states (e.g., velocity, thrust, staging events). Uncertainties in vehicle and control system characteristics and wind variability are used to define three-sigma limits to the trajectory profile. The nominal and three-sigma limits are used as references during launch and are depicted on the Range Safety Display System. These data, which define the baseline path for the vehicle, are essential to any safety study. The data are also necessary during launch because deviation from the nominal trajectory may indicate a dangerous failure.

Vehicle component reliability. The launch customer provides estimates of component and subcomponent reliability to range safety personnel. These reliabilities are generally computed using fault tree analyses. If operational experience is available, component reliabilities may be adjusted based on observed failure rates. The adjustment process uses conventional filtering theory for estimating the confidence level for operational and estimated reliability but also includes a degree of subjectivity and technical judgment.

The use of fault trees to estimate system reliability is quite common in risk management. Fault tree analysis is most effective when subcomponent reliabilities are well known (e.g., through repeated laboratory tests) but may be less accurate in estimating reliabilities when failure modes are dependent or unexpected. Adding complexity to a fault tree (e.g., adding nodes) does not necessarily result in a more accurate estimate of reliability because the uncertainties in each component propagate throughout the tree.

Vehicle failure modes, probabilities, and effects. Probable failure modes are identified by the launch customer using event trees and component reliabilities. This process includes describing each failure type (including the results of command destruct), its likelihood as a function of time, its effect on the vehicle's trajectory (e.g., a change in thrust direction), and the quantity, type, and energy of debris that would be generated. These data may also be adjusted by range safety personnel based on previous experience.

Wind modeling and debris-dispersion modeling. Statistics on monthly or seasonal winds are developed at each range to determine the likely trajectories of expended stages or debris. These data include the average wind magnitude and direction as a function of altitude, as well as the statistical variability of these parameters. Wind speed or direction shifts downrange are not considered.

At the time of launch, the actual measured winds from aerial soundings may be used to improve prelaunch estimates. The wind data are used with the data on ballistic coefficient and energy to determine debris trajectories. During launch, wind and aerodynamic effects are omitted when computing the instantaneous impact point (IIP), but measured winds are used to depict probable debris impact points on the Range Safety Display System.

Population modeling. Simplified models of population density are developed by the ranges to determine the likelihood of casualties if debris lands in a given region. These models generally break the landmasses into regions in which the population is assumed to be equally distributed. Dense population centers and cities are separated from rural areas. Population data are available in the models for much of the world, although data for some regions, including Europe, are missing. Different population distributions and shelter probabilities are assigned depending on the time of launch (day, evening, or night).

Debris-effect modeling. Data relating object energy and the likelihood that an object will cause injuries or deaths are used to determine the smallest objects that should be included in subsequent analyses. This modeling considers the type of shelter available and the probability that a fragment of a given energy would penetrate the shelter. This analysis is also used to determine the minimum size of debris that could endanger aircraft and ships.

Computation and application of safety metrics. Safety metrics, such as casualty expectation (E_c) and the individual hit probability for aircraft or ships (P_i,) are calculated throughout the launch trajectory by computing the probability of failure at any given time; determining the potential failure modes, debris types, and energies; propagating the debris using wind and aerodynamic models; and estimating casualties for the debris type and energy, the affected area, shelter types, and population densities.

The Western Range (WR) uses the Launch Risk Analysis (LARA) computer program, along with several other analysis tools, to calculate safety metrics. Thrust termination, on-trajectory breakup, and malfunction turns are the primary failure modes considered in the LARA analysis. The Eastern Range (ER) uses a different computer program, DAMP (facility DAMage and Personnel injury), along with other packages, such as RAFIP (Random Attitude Failure Impact Predictions), RSTT (Range Safety Tumble Turns), and DISP (impact DISPersions). DAMP considers six failure modes: explosion on the launch pad, loss of control at liftoff, straight-up flight, on-trajectory failure, malfunction turn, and planned jettison of components.

The overall approaches used by the WR and ER are similar in terms of failure modeling, debris propagation, and casualty estimation. The assumptions and implementation of these methods, however, are different. RAFIP assumes that an instantaneous turn to any attitude is possible, whereas LARA uses physical limitations on turn rates. Both approaches are conservative. Conservatism is further increased by RAFIP, which assumes that no debris is consumed by heat during reentry and that no populations are sheltered. The conservatism of safety metrics computed by LARA is increased by the use of unrealistically high failure rates.

Some sensitivity analyses have been performed to determine how E_c varies with changes in input parameters, such as overall probability of failure, residual thrust, or roof protection. These sensitivity analyses identify parameters with the largest impact on the value of E_c and, therefore, show where accuracy is most important. This information can be useful for improving risk analysis methods.

Flight hazard and flight caution area. The sizes of flight hazard and caution areas are based on estimates of risk to unsheltered personnel. These areas are conservatively defined using worst-case wind conditions and a probability of vehicle failure of 1.

Blast-effect modeling. Blast risks are estimated using two tools, GLASSC, which relates blast overpressure to window breakage and casualties, and BLASTC (at the WR) or BLASTX (at the ER), which use wind and temperature profiles to determine the risk of casualties.[1] The models produce series of predicted overpressure contours and risk profiles (plots of the probability of varying numbers of casualties, assuming that the probability of vehicle failure is 1.

Toxic-effect modeling. The risks from toxic gases are estimated using two software packages. The Rocket Exhaust Effluent Diffusion Model (REEDM) predicts the toxic chemical concentration in the event of a vehicle failure and produces contours showing the predicted concentrations of toxic chemicals near the ground. The Launch Area Toxic Risk Assessment (LATRA) program is used at the WR (and will be used at the ER in the near future) to determine the likelihood of an accident, estimate individual and collective risk (P_c and E_c), and develop risk profiles based on current weather conditions, models of population density and sheltering, and the amount, type, and toxicity of the substances that could be released. Both blast and toxic risk evaluations are performed well before each launch using statistical wind conditions, and they are repeated on launch day using measured winds.

Impact limit lines. Impact limit lines (ILLs), which are defined using geographic references, define boundaries beyond which significant pieces of debris should not penetrate. The definition of ILLs does not explicitly take safety metrics into consideration; rather, it is based on preventing the overflight of inhabited landmasses whenever possible.

Instantaneous impact point. To monitor the vehicle's progress relative to the nominal trajectory and the ILLs, the vehicle's current position and instantaneous impact point (IIP) are computed and displayed in real time during flight. For computational efficiency, the vacuum IIP is used (i.e., calculations do not include aerodynamic effects).

Destruct lines. Destruct lines, located inside the ILLs, are used to ensure that significant amounts of debris will not cross the ILL. The IIP position relative to the destruct line is a primary element of information in destruct decisions during launch. Small debris will propagate farther than large debris but is generally less dangerous upon impact. Ignoring small pieces of debris results in a wider launch corridor and reduces the probability that a mission will be aborted unnecessarily.

Collision avoidance. The intended launch trajectory is compared with the trajectories of satellites in orbit that are manned or capable of being manned. If a vehicle is projected to pass within 200 km of a satellite, the launch window is adjusted. A buffer of two to eight minutes is added to the window to account for uncertainties in the accuracy

---

[1]GLASSC, BLASTC, and BLASTX are descriptive nicknames, not acronyms.

and timing of the trajectory. Because the spatial buffer is so large, this safety requirement may be quite conservative.

ACTIVITIES DURING LAUNCH

Some of the information displayed and used to make safety-related decisions during launch is different at the ER and WR. The primary tools and procedures that are common to both ranges are described below, followed by a description of methods used by just one.

Methods Common to the Western and Eastern Ranges

Both the WR and ER use a range safety display system that provides a real-time depiction of the vehicle's current position relative to the nominal trajectory. The display also shows the three-sigma dispersions around the nominal trajectory, the IIP, destruct lines, ILLs, and geographic features, such as coastlines. The map may be manually or automatically scaled as the vehicle progresses along its trajectory. The mission flight control officer (MFCO) also has a vertical display (specific to each range, as described below) and flight termination system (FTS) arm and destruct buttons on a console.

Methods Specific to the Western Range

LARA is rerun approximately two hours before launch to identify any changes in E_c caused by current wind data. The results are briefed to the MFCO and range commander. A debris pattern footprint is displayed on the range safety display system showing the probable (two-sigma) locations of debris for several postulated failure conditions. The display is updated in real time during flight. The footprints are shown as circles rather than ellipses to simplify computation.

Two specific times of interest are computed and displayed to the MFCO. Amber time is the time at which the launch vehicle has enough energy to impact a region outside the ILLs. If tracking of the vehicle is not be available by amber time, the flight is terminated. Computations for amber time are conservative in that they do not account for aerodynamic effects on the vehicle and assume the worst-case trajectory toward the ILL. MFCO response time is not included in the calculation because the MFCO is expected to be monitoring the situation closely. Red time is the time at which a straight-up vehicle would present a danger. Red time is calculated using statistical wind conditions and MFCO reaction time. If a vehicle fails to initiate its pitch program (turn downrange) by red time, the flight is terminated.

The MFCO also has a display of two vertical planes. One is used to determine whether the vehicle is pitching correctly downrange. The other shows the vehicle's cross-track position relative to destruct lines.

Methods Specific to the Eastern Range

On launch day, the measured wind profile is compared with the previously developed maximum-wind constraints. Winds in excess of these values may result in a launch hold because E_c could be increased beyond the accepted standard.

The MFCO uses two vertical profile displays to monitor the vehicle relative to the nominal trajectory, ILLs, and destruct lines. A straight-up time (analogous to red time at the WR) is also computed and displayed for reference.

At the ER, a "chevron line" display, which is designed to protect the region behind the launch site from a vehicle that does not pitch downrange successfully, is also provided. The display shows destruct lines that move downrange in real time in response to the vehicle's velocity. If the vehicle is not progressing downrange as expected, the flight is terminated before the point at which debris would pass beyond the ILLs. Generally, the chevron display is only needed for the first 100 seconds of flight.

Acronyms

AFMC	Air Force Materiel Command	MOA	memorandum of agreement
AFSPC	Air Force Space Command		
ATC	air traffic control	NASA	National Aeronautics and Space Administration
		NOTAM	Notice to Airmen
CRaSS	Cape Canaveral Range Surveillance System	NRC	National Research Council
DoD	Department of Defense	P_c	casualty probability (individual risk standard)
		P_i	individual hit probability (for aircraft or ships)
E_c	casualty expectation (collective risk standard)		
ER	Eastern Range	RCC	Range Commanders Council
		RLV	reusable launch vehicle
FAA	Federal Aviation Administration	ROCC	Range Operations Control Center
FTS	flight termination system	RSA	Range Standardization and Automation (program)
GPS	global positioning system		
		SLBM	submarine launched ballistic missile
ICBM	intercontinental ballistic missile	SMC	Space and Missile Systems Center
IIP	instantaneous impact point		
ILL	impact limit line	TMIG	telemetered inertial guidance
IMU	inertial measurement unit		
IPT	integrated product team	VFR	visual flight rules
MFCO	mission flight control officer	WR	Western Range